Sex Hormone Receptor Signals in Human Malignancies

Sex Hormone Receptor Signals in Human Malignancies

Special Issue Editor

Hiroshi Miyamoto

MDPI • Basel • Beijing • Wuhan • Barcelona • Belgrade

Special Issue Editor
Hiroshi Miyamoto
University of Rochester
USA

Editorial Office
MDPI
St. Alban-Anlage 66
4052 Basel, Switzerland

This is a reprint of articles from the Special Issue published online in the open access journal *International Journal of Molecular Sciences* (ISSN 1422-0067) from 2018 to 2019 (available at: https://www.mdpi.com/journal/ijms/special_issues/sex_hormone_receptor)

For citation purposes, cite each article independently as indicated on the article page online and as indicated below:

LastName, A.A.; LastName, B.B.; LastName, C.C. Article Title. *Journal Name* **Year**, *Article Number,* Page Range.

ISBN 978-3-03921-173-9 (Pbk)
ISBN 978-3-03921-174-6 (PDF)

Contents

About the Special Issue Editor

Hiroshi Miyamoto MD, PhD, completed his medical school and urology residency training, followed by clinical urology practice and translational research in genitourinary cancers at Yokohama City University School of Medicine and affiliated hospitals in Japan. In 1996, he moved to the United States to conduct postdoctoral research at University of Wisconsin–Madison and University of Rochester. He then completed residency training in anatomic pathology at University of Rochester Medical Center and clinical fellowship in urologic pathology at The Johns Hopkins Hospital. Since 2019, he has been the faculty as a surgical pathologist as well as an independent investigator at University of Rochester School of Medicine and Dentistry (2009–2013 and 2016–present) and Johns Hopkins University School of Medicine (2013–2016). He is currently the Professor of Pathology and Laboratory Medicine, Urology, and Oncology, and also serves as the Director of Genitourinary Pathology at University of Rochester Medical Center. He has published more than 200 peer-reviewed articles and 17 book chapters, and edited the book "Gordo's Guide to GU Pathology: A Resource for Urology and Pathology Residents". In addition, he has served as the Editor-in-Chief of *Integrative Cancer Science and Therapeutics* and as an Editorial Board member of 10 other journals, including *Pathology International, Medicine, International Journal of Molecular Sciences, and Cells.*

Preface to "Sex Hormone Receptor Signals in Human Malignancies"

Sex steroids, including androgens, estrogens, and progestogens, are known to have widespread physiological actions beyond the reproductive system via binding to the sex hormone receptors, members of the nuclear receptor superfamily that function as ligand-inducible transcription factors. Meanwhile, emerging evidence has indicated that sex hormone receptor-mediated signals are involved in the development and progression of some malignancies, such as prostate and breast carcinomas, as well as others that have not traditionally been considered as endocrine-related neoplasms. This Special Issue "Sex Hormone Receptor Signals in Human Malignancies" aims to cover a variety of aspects of the potential role of sex hormone receptor-mediated signals in prostate cancer, breast cancer, and other neoplastic conditions. The current observations described may provide unique insights into novel or known functions of sex hormone receptors and related molecules.

Hiroshi Miyamoto
Special Issue Editor

International Journal of
Molecular Sciences

Editorial

Sex Hormone Receptor Signals in Human Malignancies

Hiroshi Miyamoto [1,2,3,*]

1   Department of Pathology & Laboratory Medicine, University of Rochester Medical Center, Rochester, NY 14642, USA
2   Department of Urology, University of Rochester Medical Center, Rochester, NY 14642, USA
3   James P. Wilmot Cancer Institute, University of Rochester Medical Center, Rochester, NY 14642, USA

Received: 29 May 2019; Accepted: 30 May 2019; Published: 31 May 2019

Sex steroids, including androgens, estrogens, and progestogens, are known to have widespread physiological actions beyond the reproductive system via binding to the sex hormone receptors, members of the nuclear receptor superfamily that function as ligand-inducible transcription factors. Meanwhile, emerging evidence has indicated the involvement of sex hormone receptor signals in the outgrowth of some malignancies, such as prostate and breast carcinomas, as well as others that have not traditionally been considered as endocrine neoplasms. This Special Issue "Sex Hormone Receptor Signals in Human Malignancies" covers various aspects of the potential role of sex hormone receptors and related signals in prostate cancer [1,2], breast cancer [3–5], and other neoplastic conditions [6–9] by depicting promising findings derived from in vitro and in vivo experiments as well as analyses of surgical specimens.

Capaia et al. [1] investigated the functional role of heterogeneous nuclear ribonucleoprotein K (HNRPK) in androgen-sensitive and castration-resistant prostate cancer cells. Their in vitro data suggested that HNRPK could induce androgen receptor (AR) transactivation and activate downstream targets via functioning as its transcriptional co-regulator. Furthermore, using a co-immunoprecipitation assay coupled with mass spectrometry, they identified several proteins that could interact with HNRPK, as well as AR, and potentially modulated sensitivity to androgen deprivation therapy in prostate cancer. Similarly, Yun et al. [2] assessed the functional role of a BRCA1-interacting protein, COBRA1, in androgen-sensitive and castration-resistant prostate cancer cells. First, COBRA1 expression in prostate cancer was found to correlate with its aggressiveness. In vitro studies then indicated that COBRA1 contributed to promoting cell growth via activating the AR. Moreover, a potent estrogen, 2-methoxyestradiol, was shown to inhibit the growth of even AR-negative DU145 cells, together with down-regulation of COBRA1 expression. These observations may offer potential therapeutic approaches for both androgen-sensitive and castration-resistant prostate cancers via targeting HNRPK and COBRA1.

Forkhead box A1 (FOXA1), as a pioneer factor that modulates the activity of AR and estrogen receptor (ER)-α, has been implicated in the development and progression of prostate and breast cancers [10]. Using high throughput chemical screening and mass spectrometry, Wang et al. [3] identified proteins that could control FOXA1 in breast cancer cells. Of these, cyclin-dependent kinase 1 was suggested to directly regulate FOXA1 via its phosphorylation. Lopez et al. [4] examined the mutational signatures of ER-positive/progesterone receptor (PR)-negative breast cancers and compared the molecular landscapes of PR-negative versus PR-positive tumors. Mutations in the *PIK3CA* (37%) and *TP53* (33%) genes were most frequently seen in PR-negative tumors, with lower (*PIK3CA*: vs. 47%, $p < 0.01$) or higher (*TP53*: vs. 19%, $p < 0.01$) prevalence compared with PR-positive tumors. Additionally, in patients with ER-positive/PR-negative breast cancer, mutations in the *PIK3CA* and/or *TP53* were found to correlate with a significantly worse prognosis. Meanwhile, Hsu et al. [5] summarized available data indicating the involvement of a putative membrane ER, G protein-coupled

ER (GPER; also known as GPR30), in breast cancer. Current evidence suggests that GPER plays an important role in mediating the genomic and non-genomic effects of estrogens in breast cancer cells. GPER expression was also suggested to serve as a prognosticator in patients with breast cancer.

Aquino et al. [6] immunohistochemically stained for AR, ERα, ERβ, GPR30, and PR in salivary gland tumor specimens. AR, ERβ, and GPR30 were positive in 25%, 36% (nuclear)/28% (cytoplasmic), and 18% (nuclear)/85% (cytoplasmic) of tumors, respectively, while ERα and PR were negative in all cases examined. In addition, there was a trend to correlate between cytoplasmic ERβ expression and higher grade ($p = 0.052$) or between nuclear GPR30 expression and better disease-free survival ($p = 0.055$). We also used immunohistochemistry to assess the expression status of phospho-ELK1, an activated form of a transcription factor ELK1, in upper urinary tract urothelial carcinoma specimens [7]. Phospho-ELK1 expression was up-regulated in tumors (47.5%; $p = 0.002$), compared with non-neoplastic urothelial tissues (25.3%), and muscle-invasive tumors (54.8%; $p = 0.065$), compared with non-muscle-invasive tumors (35.1%), and was associated with risks of disease progression ($p = 0.055$) and cancer-specific mortality ($p = 0.008$). More interestingly, phospho-ELK1 expression in tumors tended to correlate with AR positivity ($p = 0.091$), especially in male patients ($p = 0.058$). These data support our previous findings in preclinical models [11–13] indicating that ELK1 induces urothelial carcinogenesis and cancer growth via cooperation with AR signaling. Another immunohistochemical study by Czogalla et al. [8] determined the expression of ERα and a transcription factor NRF2, which was shown to physically interact with ERα [14], in ovarian cancer tissue samples. The levels of cytoplasmic NRF2 expression were significantly higher in low grade tumors than in high grade tumors ($p = 0.03$). In addition, patients with NRF2-high ($p = 0.04$) or ERα-high ($p = 0.002$) serous cancer showed significantly better overall survival. As expected, inactivation of NRF2 (i.e. cytoplasmic expression in tissues, siRNA expression in cell lines) resulted in up-regulation of ERα protein/mRNA expression, supporting the crosstalk between NRF2 and ERα in ovarian cancer cells. Finally, Coricovac et al. [9] assessed the cytotoxic effects of the components of oral contraceptives in normal skin and skin cancer cells. Ethinylestradiol (10 μM), levonorgestrel (10 μM), or both inhibited the growth of all cell lines examined, especially melanoma cells. However, conflicting results on the effects of contraceptives on the viability of melanoma cells with UVB irradiation were obtained: additional inhibition (in human A375 line) vs. protection against UVB-induced suppression (in murine B164A5 line). Further studies are thus warranted to determine the impact of hormonal therapy with or without irradiation on skin tumorigenesis and tumor progression.

Again, a variety of aspects of the role of sex hormone receptor-mediated signals in human malignancies are described in this Special Issue. The current observations may thus provide a unique insight into novel or known functions of sex hormone receptors and related molecules.

Conflicts of Interest: The author declares no competing interest.

References

1. Capaia, M.; Granata, I.; Guarracino, M.; Petretto, A.; Inglese, E.; Cattrini, C.; Ferrari, N.; Boccardo, F.; Barboro, P. A hnRNP K–AR-related signature reflects progression toward castration-resistant prostate cancer. *Int. J. Mol. Sci.* **2018**, *19*, 1920. [CrossRef] [PubMed]
2. Yun, H.; Bedolla, R.; Horning, A.; Li, R.; Chiang, H.C.; Huang, T.H.; Reddick, R.; Olumi, A.F.; Ghosh, R.; Kumar, A.P. BRCA1 interacting protein COBRA1 facilitates adaptation to castrate-resistant growth conditions. *Int. J. Mol. Sci.* **2018**, *19*, 2104. [CrossRef] [PubMed]
3. Wang, S.; Singh, S.K.; Katika, M.R.; Lopez-Aviles, S.; Hurtado, A. High throughput chemical screening reveals multiple regulatory proteins on FOXA1 in breast cancer cell lines. *Int. J. Mol. Sci.* **2018**, *19*, 4123. [CrossRef] [PubMed]
4. Lopez, G.; Costanza, J.; Colleoni, M.; Fontana, L.; Ferrero, S.; Miozzo, M.; Fusco, N. Molecular insights into the classification of luminal breast cancers: The genomic heterogeneity of progesterone-negative tumors. *Int. J. Mol. Sci.* **2019**, *20*, 510. [CrossRef] [PubMed]
5. Hsu, L.H.; Chu, N.M.; Lin, Y.F.; Kao, S.H. G-protein coupled estrogen receptor in breast cancer. *Int. J. Mol. Sci.* **2019**, *20*, 306. [CrossRef] [PubMed]

6. Aquino, G.; Collina, F.; Sabatino, R.; Cerrone, M.; Longo, F.; Ionna, F.; Losito, N.S.; De Cecio, R.; Cantile, M.; Pannone, G.; et al. Sex hormone receptors in benign and malignant salivary gland tumors: Prognostic and predictive role. *Int. J. Mol. Sci.* **2018**, *19*, 399. [CrossRef] [PubMed]
7. Inoue, S.; Ide, H.; Fujita, K.; Mizushima, T.; Jiang, G.; Kawahara, T.; Yamaguchi, S.; Fushimi, H.; Nonomura, N.; Miyamoto, H. Expression of phospho-ELK1 and its prognostic significance in urothelial carcinoma of the upper urinary tract. *Int. J. Mol. Sci.* **2018**, *19*, 777. [CrossRef] [PubMed]
8. Czogalla, B.; Kahaly, M.; Mayr, D.; Schmoeckel, E.; Niesler, B.; Kolben, T.; Burges, A.; Mahner, S.; Jeschke, U.; Trillsch, F. Interaction of ERα and NRF2 impacts survival in ovarian cancer patients. *Int. J. Mol. Sci.* **2019**, *20*, 112. [CrossRef] [PubMed]
9. Coricovac, D.; Farcas, C.; Nica, C.; Pinzaru, I.; Simu, S.; Stoian, D.; Soica, C.; Proks, M.; Avram, S.; Navolan, D.; et al. Ethinylestradiol and levonorgestrel as active agents in normal skin, and pathological conditions induced by UVB exposure: In vitro and in ovo assessments. *Int. J. Mol. Sci.* **2018**, *19*, 3600. [CrossRef] [PubMed]
10. Augello, M.A.; Hickey, T.E.; Knudsen, K.E. FOXA1: master of steroid receptor function in cancer. *EMBO J.* **2011**, *30*, 3885–3894. [CrossRef] [PubMed]
11. Kawahara, T.; Ide, H.; Kashiwagi, E.; Patterson, J.D.; Inoue, S.; Shareef, H.K.; Aljarah, A.K.; Zheng, Y.; Baras, A.S.; Miyamoto, H. Silodosin inhibits the growth of bladder cancer cells and enhances the cytotoxic activity of cisplatin via ELK1 inactivation. *Am. J. Cancer Res.* **2015**, *5*, 2959–2968. [PubMed]
12. Kawahara, T.; Shareef, H.K.; Aljarah, A.K.; Ide, H.; Li, Y.; Kashiwagi, E.; Netto, G.J.; Zheng, Y.; Miyamoto, H. ELK1 is up-regulated by androgen in bladder cancer cells and promotes tumor progression. *Oncotarget* **2015**, *6*, 29860–29876. [CrossRef] [PubMed]
13. Inoue, S.; Ide, H.; Mizushima, T.; Jiang, G.; Kawahara, T.; Miyamoto, H. ELK1 promotes urothelial tumorigenesis in the presence of an activated androgen receptor. *Am. J. Cancer Res.* **2018**, *8*, 2325–2336. [PubMed]
14. Ansell, P.J.; Lo, S.C.; Newton, L.G.; Epinosa-Nicholas, C.; Zhang, D.D.; Liu, J.H.; Hannink, M.; Lubahn, D.B. Repression of cancer protective genes by 17β-estradiol: Ligand-dependent interaction between human Nrf2 and estrogen receptor α. *Mol. Cell. Endocrinol.* **2005**, *243*, 27–34. [CrossRef] [PubMed]

International Journal of
Molecular Sciences

Article

A hnRNP K–AR-Related Signature Reflects Progression toward Castration-Resistant Prostate Cancer

Matteo Capaia [1,†], Ilaria Granata [2,†], Mario Guarracino [2], Andrea Petretto [3], Elvira Inglese [3], Carlo Cattrini [1,4], Nicoletta Ferrari [5], Francesco Boccardo [1,4] and Paola Barboro [1,*]

[1]   Academic Unit of Medical Oncology, Ospedale Policlinico San Martino-IRCCS, L.go R. Benzi 10, 16132 Genova, Italy; matteo.capaia@hsanmartino.it (M.C.); carlo.cattrini@gmail.com (C.C.); fboccardo@unige.it (F.B.)

[2]   Institute for High Performance Computing and Networking (ICAR), National Research Council (CNR), Via Pietro Castellino 111, 80131 Napoli, Italy; ilaria.granata@icar.cnr.it (I.G.); mario.guarracino@cnr.it (M.G.)

[3]   Core Facilities-Proteomics Laboratory, Giannina Gaslini Institute, L.go G. Gaslini 5, 16147 Genova, Italy; AndreaPetretto@gaslini.org (A.P.); elvira.inglese@gmail.com (E.I.)

[4]   Department of Internal Medicine and Medical Specialties, School of Medicine, University of Genova, L.go R. Benzi 10, 16132 Genova, Italy

[5]   Molecular Oncology and Angiogenesis, Ospedale Policlinico San Martino-IRCCS, L.go R. Benzi 10, 16132 Genova, Italy; nicoletta.ferrari@hsanmartino.it

[*]   Correspondence: paola.barboro@hsanmartino.it; Tel.: +39-010-5558-539; Fax: +39-010-5558-368

[†]   These authors contributed equally to this work.

Received: 15 June 2018; Accepted: 29 June 2018; Published: 30 June 2018

Abstract: The major challenge in castration-resistant prostate cancer (CRPC) remains the ability to predict the clinical responses to improve patient selection for appropriate treatments. The finding that androgen deprivation therapy (ADT) induces alterations in the androgen receptor (AR) transcriptional program by AR coregulators activity in a context-dependent manner, offers the opportunity for identifying signatures discriminating different clinical states of prostate cancer (PCa) progression. Gel electrophoretic analyses combined with western blot showed that, in androgen-dependent PCa and CRPC in vitro models, the subcellular distribution of spliced and serine-phosphorylated heterogeneous nuclear ribonucleoprotein K (hnRNP K) isoforms can be associated with different AR activities. Using mass spectrometry and bioinformatic analyses, we showed that the protein sets of androgen-dependent (LNCaP) and ADT-resistant cell lines (PDB and MDB) co-immunoprecipitated with hnRNP K varied depending on the cell type, unravelling a dynamic relationship between hnRNP K and AR during PCa progression to CRPC. By comparing the interactome of LNCaP, PDB, and MDB cell lines, we identified 51 proteins differentially interacting with hnRNP K, among which KLK3, SORD, SPON2, IMPDH2, ACTN4, ATP1B1, HSPB1, and KHDRBS1 were associated with AR and differentially expressed in normal and tumor human prostate tissues. This hnRNP K–AR-related signature, associated with androgen sensitivity and PCa progression, may help clinicians to better manage patients with CRPC.

Keywords: castration-resistant prostate cancer; heterogeneous nuclear ribonucleoprotein K; androgen receptor; androgen deprivation therapy

---

1. Introduction

In the last years, experimental evidence has supported the role of the androgen receptor (AR) in castration-resistant prostate cancer (CRPC) development. Almost all patients with metastatic prostate cancer (PCa) initially treated with androgen deprivation therapy (ADT) progress to CRPC, in many cases following the reactivation of the AR pathway. Several mechanisms, which are not necessarily mutually exclusive, have been proposed to explain CRPC development [1]. They include AR amplification or overexpression, AR mutations that can modify the ligand specificity, AR gain of function, changes in the expression levels of AR coregulators, and involvement of alternative pathways that could be completely independent of AR signaling [2,3]. However, to date, the molecular mechanisms through which hormone-sensitive PCa cells acquire the ability to resist to hormone deprivation need further investigation.

Multi-omics studies showed that CRPC is a heterogeneous group of diseases characterized by different genotypes and phenotypes [4–6]. Emerging evidence suggests that phenotypic plasticity, driving the adaptation to ADT stress, further amplifies cellular heterogeneity and can contribute to ADT resistance [7,8]. These findings indicate that an adequate response to cytotoxic or targeted therapies cannot disregard the broad spectrum of cellular sub-clones with different clinical behaviors. Since multiple pathways are involved in CRPC development and progression, it is evident that therapies, to be useful, should selectively target driving molecular alterations at a specific stage of PCa evolution. For this purpose, we explored the possibility of developing a signature for identifying PCa and CRPC subtypes with different androgen responsiveness.

A new scenario to develop alternative targeted therapies was recently proposed by Liu et al. [9], reporting the context dependency by which AR coregulators control selectively an AR target gene set reflecting PCa biology and evolution. This finding also provides a proof of principle for the identification of PCa subtypes associated with an AR coregulator and the corresponding subset of AR-related genes.

The heterogeneous nuclear ribonucleoprotein K (hnRNP K) is a multifunctional protein playing a pivotal role in regulating numerous cellular functions, such as transcription, signal transduction, alternative splicing, and chromatin remodeling [10]. This functional versatility depends on post-translation modifications (PTM) that modulate its interactions with nucleic acids and proteins [11]. Increasing evidence for the involvement of hnRNP K in cancer progression was reported [12]. In PCa patients, we have demonstrated that its overexpression positively correlates with Gleason score and poor patients prognosis [13] and that the concomitant expression of both AR and cytoplasmic hnRNP K has a potential prognostic value [14]. In PCa cell lines, hnRNP K regulates AR activity by inhibiting its translation [15], and, in nucleoplasm, hnRNP K phosphorylation shapes the AR–DNA complex after anti-androgen treatments [16]. HnRNP K appears to be also able to regulate neuroendocrine differentiation [17].

Kelly et al. [18] underlined the importance of the adaptive phenotype acquired during ADT leading to cellular reprogramming that ultimately resulted in tumor heterogeneity and different AR status. To reproduce this behavior, we obtained the androgen-resistant cell lines PDB and MDB by treating LNCaP cell line for over one year with the anti-androgen bicalutamide (BIC) in the presence or absence of 5-α-dihydrotestosterone (DHT), respectively. Transcriptomic and proteomic analyses highlighted the high degree of phenotypic plasticity that characterizes our CRPC models and allows their adaptation under stress conditions. BIC-resistant cell lines represent two sub-populations with AR levels and subcellular localization similar to the parental LNCaP cell line, but with reduced functionality depending on AR phosphorylation status. Interestingly, partial (PDB) or minimal (MDB) AR transcriptional activity correlated with enhanced tumorigenicity and decreased sensitivity to treatment with a novel anti-androgen, enzalutamide, compared to the parental cell line [19].

Given the above findings, in this study, we investigated the role of hnRNP K in androgen-resistance using in vitro models of androgen-dependent or castration-resistant PCa. We hypothesized that hnRNP K, working like an AR transcriptional collaborator, could participate in regulating the different AR transcriptional programs during PCa development and progression. Consequently, a hnRNP K signature associated to AR activity could be useful to identify clinically distinct PCa subgroups.

2. Results

2.1. Role of hnRNP K in ADT Resistance

We previously reported that in the androgen-dependent cell line LNCaP, changes in the AR binding property of hnRNP K was associated with cell growth and AR activity [16,20]. Here, using our in vitro resistant models, we investigated the role of hnRNP K in the resistant cell lines PDB and MDB which can be considered models mimicking two CRPC subpopulations [19].

HnRNP K silencing decreased both AR and PSA expression in LNCaP cells (Figure 1a,b), while it was less effective in BIC-resistant cells lines, in particular MDB. Furthermore, hnRNP K silencing in LNCaP induced a 68% reduction of AR activity, as evaluated by the luciferase assay (Figure 1c). Because of their reduced AR activity [19], it was not possible to determine luciferase activity in PDB and MDB cells.

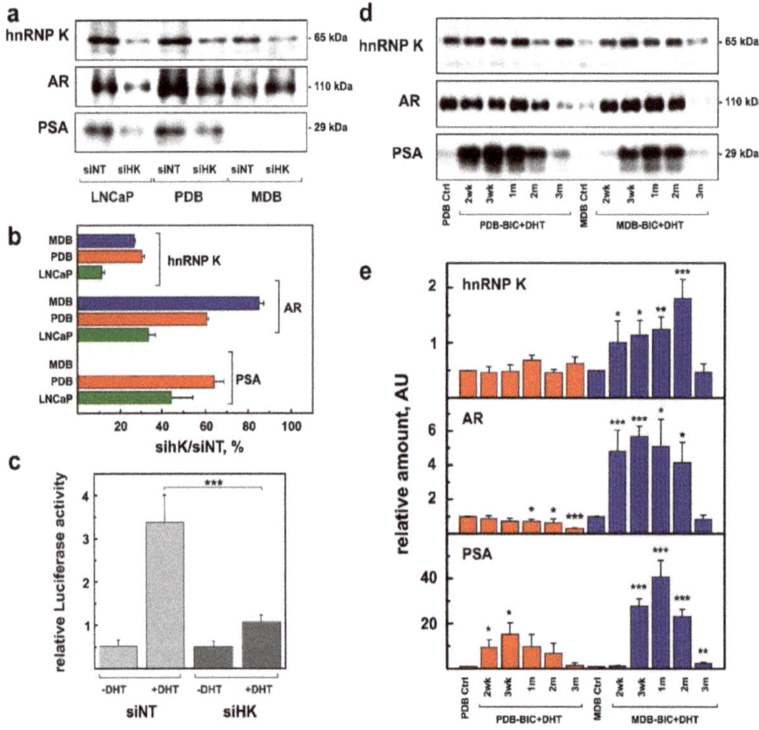

Figure 1. Effects of heterogeneous nuclear ribonucleoprotein K (hnRNP K) on the regulation of androgen receptor (AR) expression and AR transcriptional activity in LNCaP, PDB, and MDB cells. (**a**,**d**) Representative western blotting (WB) analysis carried out using antibodies against hnRNP K, AR, and prostate-specific antigen (PSA) and (**b**,**e**) quantitative analysis of total proteins extract from (**a**) hnRNP K-silenced (siHK) and -non-silenced (siNT) LNCaP, PDB, and MDB cell lines or (**d**) from control (Ctrl) PDB and MDB cell lines grown in the appropriate medium (see Material and Methods) or in restored LNCaP growth medium without bicalutamide (-BIC) and with 5-α-dihydrotestosterone(+DHT) for 2, 3 weeks (wk) or 1, 2, 3 months (m). The ordinate represents the mean ± SE (**b**) of the percentage of selected protein expression in hnRNP K-silenced cell lines or (**e**) the relative amounts of proteins determined by quantitative analysis. (**c**) AR transcriptional activity determined in hnRNP K-silenced and non-silenced LNCaP by the luciferase activity assay; * $p < 0.05$, ** $p < 0.01$, and *** $p < 0.001$ (Student's *t*-test).

BIC removal and restored availability of androgen in the culture media determined comparable features in prostate-specific antigen (PSA) expression in both our ADT-resistant cell lines, with a maximum effect after 3–4 weeks (Figure 1d,e). However, in MDB cells, the substantial increase in PSA synthesis was associated with the overexpression of both hnRNP K and AR, while, in PDB cells, it was independent of it, probably due to AR hypersensitivity developed during ADT.

Overall, these results suggest that hnRNP K in prostate cancer cell lines is likely to act as an AR transcriptional collaborator that regulates AR activity through different molecular mechanisms depending on cell differentiation, supporting the hypothesis of the involvement of hnRNP K in ADT resistance.

2.2. Role of hnRNP K Phosphorylation in ADT Resistance

The mechanistic role of phosphorylation in regulating hnRNP K transcriptional activity [21,22] as well as its increased expression in several neoplasms with an aggressive phenotype [12] have been described. Here, we evaluated both its expression level and phosphorylation status in LNCaP and BIC-resistant PDB and MDB cell lines.

Unexpectedly, using quantitative western blotting (WB) analysis (Figure 2a), we detected a significant hnRNP K decrease in PBD and MDB cells compared to the parental cell line LNCaP, suggesting its distinct functional role in androgen-resistance compared to PCa where hnRNP K overexpression has shown diagnostic and prognostic value [13].

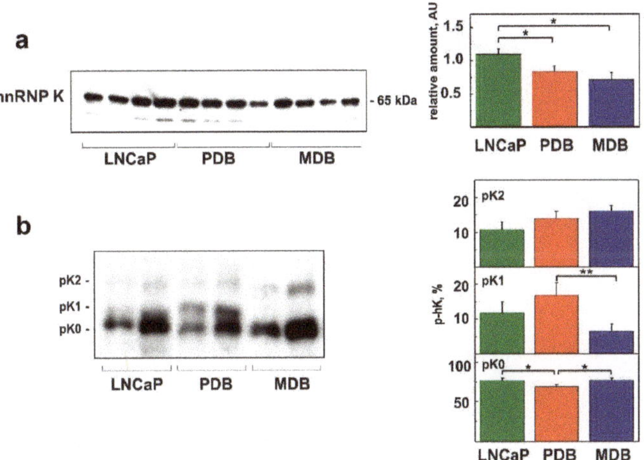

Figure 2. HnRNP K decreased expression and altered phosphorylation in resistant cell lines PDB and MDB compared with LNCaP. (**a**) WB analysis of hnRNP K expression in total extracts from LNCaP, PDB, and MDB cell lines. The histogram on the right side represents the mean ± SE of relative amount of proteins determined in six WBs. (**b**) The quantitative analysis of hnRNP K phosphorylation was carried out using 1D Phos-tag and WB analysis of LNCaP, PDB, and MDB total extracts. The histograms on the right side represent the mean percentage ± SE of phosphorylated hnRNP K (p-hK) evaluated in four experiments. The hnRNP K isoforms with minimal (phK0), intermediate (phK1), or maximal (phK2) phosphorylation are indicated; * $p < 0.05$ and ** $p < 0.01$ (Student's *t*-test).

By evaluating the total hnRNP K phosphorylation status by means of the monodimensional phosphate affinity gel electrophoresis (1D Phos-tag), it was possible to identify a non-phosphorylated isoform (phK0) and two species characterized by intermediate (phK1) and maximal (phK2) phosphorylation (Figure 2b). As shown in the histogram of Figure 2b (right panel), significant differences were found only for PDB, showing phK0 decrease, compared to LNCaP and MDB, and phK1 increase with respect to MDB.

Phosphorylation and cellular compartmentalization of hnRNP K isoforms generate a regulatory system involved in cell growth [23] and translation regulation [24]. Aberrant hnRNP K hyperphosphorylation and cytoplasmic accumulation are peculiar features of several human tumors, often associated with a worse prognosis [12]. As serine residues phosphorylation and hnRNP K functions involved in regulating its intracellular distribution, cell growth, and transcription (PhosphoSitePlus database: www.phosphosite.org) are closely related, we evaluated the role of the subcellular distribution of hnRNP K phosphorylated isoforms in PCa evolution. Using Phos-tag bidimensional gel electrophoresis (2D Phos-tag) and WB (Figure 3), we analyzed nuclear and cytoplasmic extracts from the androgen-dependent PCa cell line LNCaP and from cell lines with different CRPC phenotypes with respect to AR status and transcriptional activity: three AR-positive cell lines (PDB, hypersensitive; MDB, inactive; 22Rv1, ARv7, androgen-independent) and the AR-negative PC3 cell line [25]. Each spot detected with the anti-hnRNP K antibody was attributed, according to Kimura el al. [26], to the alternatively spliced isoforms 1 and 2 and to the phosphorylated isoforms at Ser116, Ser284, Ser353 residues (pS116, pS284, pS353), as schematically represented in Figure S1. Quantitative analysis of each spot was carried out using PDQuest software and is reported in Table 1 and Figure S2.

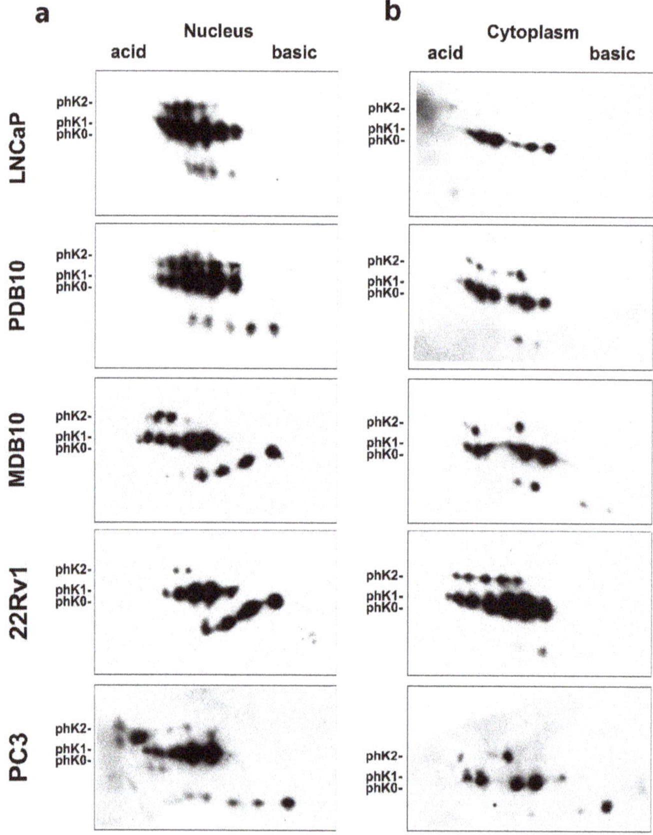

Figure 3. The hnRNP K phosphorylation status correlates with AR activity in prostate cancer cell lines. Nuclear (**a**) and cytoplasmic (**b**) profiles of hnRNP K phosphorylated isoforms in androgen-responsive LNCaP and resistant cell lines PDB, MDB, 22Rv1, and PC3, evaluated using 2D Phos-tag and WB analysis. The membranes were probed with an anti-hnRNP K antibody. The phK0, phK1, and phK2 identified in Phos-tag 1D (Figure 2b) are indicated.

Table 1. Quantitative analysis of the spots reported in Figure 3. The assignment of the alternatively spliced isoforms 1 and 2, non-phosphorylated, and S116-, S284-, S353-phosphorylated forms was carried out according to the schematic representation showed in Figure S1.

Cell Line	LNCaP(AR-FL [1])	PDB(AR-FL)	MDB(AR-FL)	22Rv1 (ARv7 [2])	PC3(AR-Null [3])
AR Status	Active	Hypersensitive	Inactive	Androgen Independent	Not Detected
Nucleus					
Alternatively spliced					
isoform 1 (%)	50.1	66.7	62.5	76.6	66.8
isoform 2 (%)	49.9	33.3	37.5	23.4	33.2
Phosphorylated forms					
pS116 (%)	17.1	19.5	15.0	1.7	0.7
pS284 (%)	37.6	45.8	23.2	36.1	22.8
pS353 (%)	21.6	8.4	16.2	1.8	10.6
Non-phosphorylated forms (%)	23.6	26.2	45.5	60.3	65.8
Cytoplasm					
Alternatively spliced					
isoform 1 (%)	38.7	45.1	66.1	63.7	76.2
isoform 2 (%)	61.3	54.9	33.9	36.3	23.8
Phosphorylated forms					
pS116 (%)	0.0	8.2	9.9	13.3	6.8
pS284 (%)	0.0	0.0	42.0	41.3	0.0
pS353 (%)	24.9	35.7	21.5	17.8	23.8
Non-phosphorylated forms (%)	75.1	56.1	26.6	27.7	69.4

[1] Cell line expressing AR full length; [2] Cell line expressing AR splicing isoform ARv7; [3] Cell line not expressing AR.

The level of nuclear hnRNP K phosphorylated isoform 1 was higher than that of isoform 2 in resistant cell lines compared to androgen-dependent LNCaP, while, in the cytoplasm, alternatively spliced isoforms phosphorylation was regulated in an opposite manner. Interestingly, pS353, localized in the nuclear shuttling (KNS) domains regulating hnRNP K intracellular localization, was detected exclusively in isoform 2 (Figure S2).

Quantitative analysis of nuclear hnRNP K phosphorylated forms in all cell lines revealed that: (i) pS353 decreased in resistant cell lines compared to LNCaP; (ii) an over 36% pS284 increase was observed in cell lines with active AR axis, while pS284 decreased to about 23% in MDB and PC3 cells; (iii) both high percentages of non-phosphorylated forms and low pS116 percentages were detected in androgen-independent cell lines MDB (AR-positive), 22Rv1 (ARv7), and PC3 (AR-negative), independently of the AR status.

The cytoplasmic hnRNP K isoforms distribution in the different cell lines showed that pS116 could discriminate between LNCaP and CRPC lines, regardless of AR functionality. Cytoplasmic pS353 showed an opposite trend in PDB and 22Rv1 cell lines with aberrant AR activation, in comparison to LNCaP. The pS284 isoforms were overrepresented in androgen-independent MDB and 22Rv1 cell lines.

These findings suggest that differential phosphorylation at specific serine residues and compartmentalization of hnRNP K splicing isoforms 1 and 2 correlate with different resistant phenotypes, providing further evidence for hnRNP K involvement in CRPC evolution.

2.3. Characterization of the hnRNP K Interactome in LNCaP, PDB, and MDB Cell Lines

HnRNP K may regulate several cellular functions, such as transcription and signal transduction, by PTMs that modify its binding partners [11,27]. Using a co-immunoprecipitation assay coupled with mass spectrometry (MS), we identified the proteins directly or indirectly interacting with hnRNP K in LNCaP, PDB, and MDB total extracts. A total of 254 proteins were identified (Table S1).

As shown in Figure 4, the number of proteins interacting with hnRNP K gradually decreased from 221 in LNCaP, to 111 in PDB, and 55 in MDB cell lines. Protein–protein interaction networks for LNCaP, PDB, and MDB cell lines were created through Intact and Reactome databases and then clusterized through the Cytoscape app clustermaker2 [28]. Three significant clusters were obtained for each cell line, and the proteins belonging to each of them were enriched through JEPETTO Cytoscape plug-in [29], according to the Gene Ontology (GO) Molecular Function annotation restricted to prostate tissue. In cluster 1, hnRNP K was the primary hub connecting about 50% of proteins both in LNCaP and in PDB cells, while cluster 2 was predominant in the resistant cell line MDB. Interestingly, some proteins involved in AR binding (Figure 5, red arrow) were significantly clustered only in the androgen-dependent cell line LNCaP. In all cell lines, enriched GO terms in cluster 1 and 2 were different, while cluster 3, referred mainly to mRNA binding, represented about 20% of hnRNP K interacting proteins. Since, in our experiments, MS analysis was not able to efficiently detect AR, we verified hnRNP K–AR association in all cell lines using WB analysis of hnRNP K co-immunoprecipitates (co-IPs) obtained from LNCaP, PDB, and MDB (Figure S3).

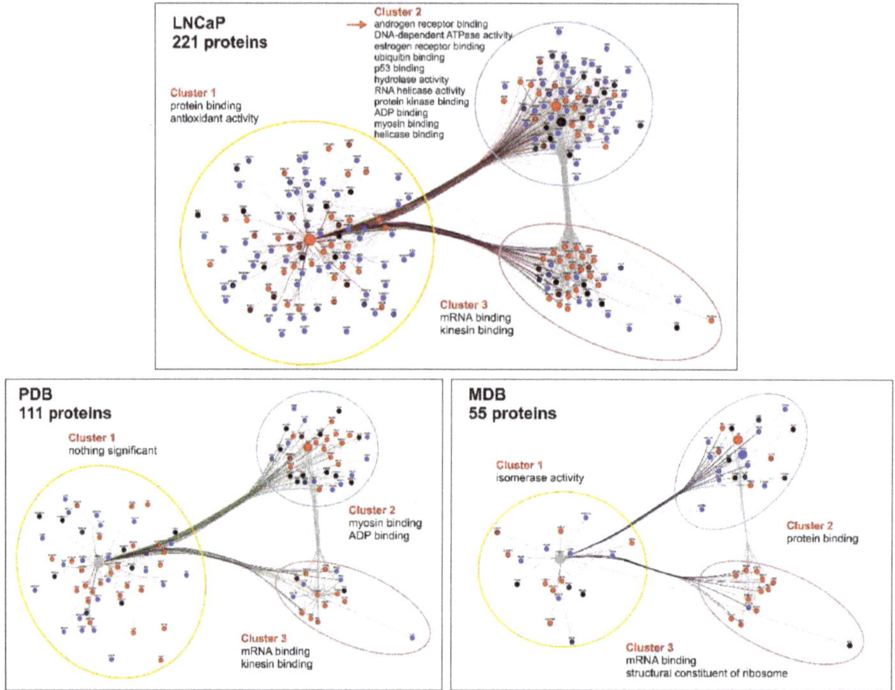

Figure 4. The hnRNP K interactome depends on the cellular phenotype. Protein–protein interaction networks of co-immunoprecipitates (co-IPs) obtained from LNCaP, PDB, and MDB, highlighting GO molecular function annotation restricted to prostate tissue grouped in three clusters, were realized in Cytoscape environment. The dot color refers to the Z score of protein intensity calculated on the basis of all three experiments: upregulated (**red**), downregulated (**blue**), and medium (**black**) value.

By comparing hnRNP K co-IP abundance values obtained from LNCaP, PDB, and MDB, we identified three sets of differentially interacting proteins. The data were linearly modeled using Limma R package, and from the three contrasts generated (PDBvsLNCaP, MDBvsLNCaP, MDBvsPDB) we identified 51 proteins differentially interacting ($p < 0.05$) with hnRNP K (DIhKP), as reported in Table S2. Decreased interaction with hnRNP K was observed in all comparisons, except for PDBvsLNCaP, where eight proteins increased their binding capacity.

Enrichment analysis was performed to identify the biological processes and molecular functions associated to DIhKPs, according to Gene Ontology annotation (Table S3). Among the significant terms, the highest percentages corresponded to proteins involved in translation (17% for GO Biological Processes) and RNA binding (35% for GO Molecular Function), indicating that in PCa resistance, the central role of hnRNP K is linked to protein synthesis regulation.

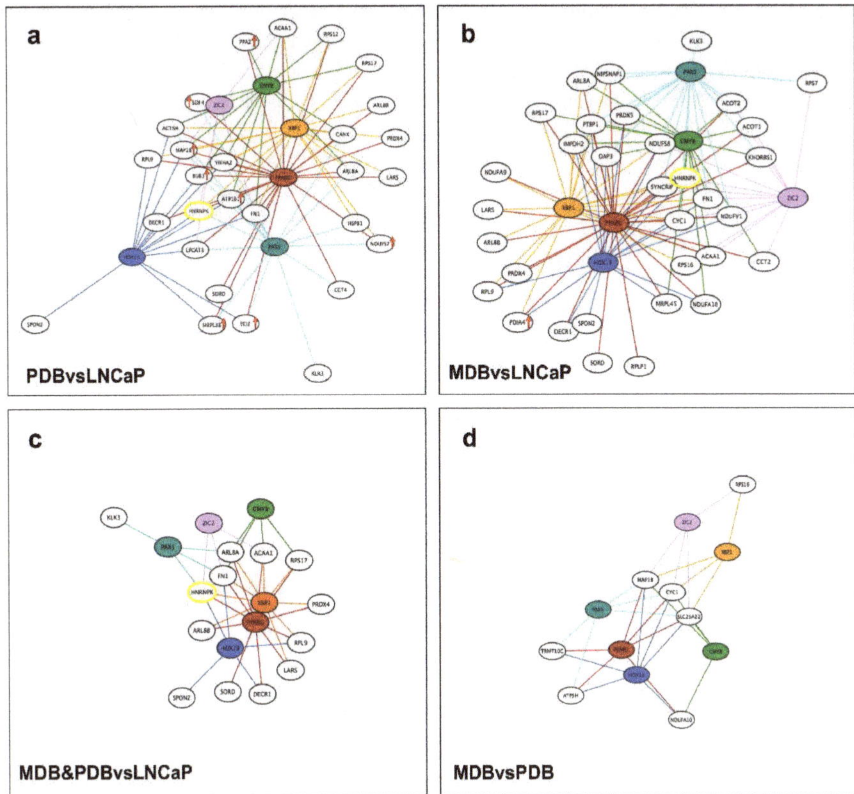

Figure 5. Transcriptional regulatory subnetworks of transcription factors (hK-TFs)potentially regulating the expression of differentially interacting hnRNP K proteins (DIhKPs) and selected DIhKPs, predicted and enriched by DAVID UCSC_TFBS option. The comparison of the subnetworks reveals dynamic remodeling of the hnRNP K interactome in LNCaP, PDB, and MDB. The six enriched hK-TFs are represented in colored circles, and DIhKPs in white circles, the edges are colored according to the hK-TFs. The subnetworks were grouped by the specific comparison of proteins abundance: PDBvsLNCaP (**a**), MDBvsLNCaP (**b**), MDB and PDBvsLNCaP (**c**), and MDBvsPDB (**d**). The red arrows indicate upregulated DIhKP expression; in all other cases, DIhKPs expression is downregulated.

In PCa and CRPC tissues, the altered profile of AR-interacting transcription factors (TFs) determines AR transcriptional program modifications [30]. By performing the enrichment of transcription factor binding sites (TFBS) through Database for Annotation Visualization and Integrated Discovery, we identified six TFs potentially regulating the expression of DIhKPs (hK-TFs) (Table S3). It is significant that five of these hK-TFs revealed a direct association with AR and PCa. ZIC2 and HOXB13 are AR-controlled genes [9]; the latter, in the absence of androgen, promotes androgen-independent growth in PCa cell line [31], while HOXA13 overexpression is significantly associated to PCa poor prognosis [32]. CMYB shares with AR common genes that correlate with advanced stages of PCa [33]. The TF XBP1, mediating cellular stress responses, is proposed as a predictor marker for PCa [34]. The bidirectional crosstalk between PPARG and AR regulates growth and development both in normal and in tumor prostate, as recently highlighted [35]. Using Cytoscape, we generated a network visualization of the predicted regulatory interaction between hK-TFs and 51 DIhKPs (Figure S4). In this network, PPARG and PAX5 were the main hK-TFs, regulating about

25% and 18% of DIhKPs, respectively. In addition, all DIhKPs and hK-TFs appeared to be potentially regulated by one or more hK-TF. Interestingly, hnRNP K expression resulted to be potentially regulated by four hK-TFs, among which PPARG, HOX13, and ZIC2 are controlled by AR.

As LNCaP, PDB and MDB cell lines can mimic distinct stages of PCa progression [19], we can infer that the Cytoscape network representation of the hK-TFs and the controlled DIhKPs grouped by comparing PDB versus LNCaP, MDB versus LNCaP, MDB and PDB versus LNCaP, and MDB versus PDB could represent proteins associated with different stages of PCa progression (Figure 5). Although PPARG represents the main hK-TF controlling DIhKPs in all networks, the global regulatory feature of hnRNP K partners is different in all comparisons. In PDB versus LNCaP, the eight proteins with increased hnRNP K interaction (Table S2) were under transcriptional control of PPARG and XBP1 (Figure 5a). Conversely, in MDB versus LNCaP, excluding PPARG, the number of DIhKPs controlled by others hK-TF was equally distributed (Figure 5b). Interestingly, despite the exiguous DIhKP number, the networks that featured the differences between androgen-dependent and -resistant cell lines or between MDB and PDB cell lines (Figure 5c,d) showed, respectively, an enhanced role of XBP1 and HOX13 in regulating hnRNP K partners transcription.

Overall, these findings indicate that, in cell lines with different androgen-resistant phenotypes, decreased hnRNP K expression and modified subcellular distribution of its isoforms were associated with dynamic remodeling of the hnRNP K interacting protein network depending on cellular differentiation, thus providing a rationale for DIhKPs analysis in human prostate cancer tissues.

2.4. Signature Identification for Potential PCa and CRPC Patient Stratification

From the above results, we hypothesized that some DIhKPs could differentiate distinct stages of PCa progression to CRPC, thus conferring clinical relevance to the findings obtained in our CRPC cell lines PDB and MDB. In this section, we describe the analytical strategies used to identify, among the 51 DIhKPs, a protein set that might putatively discriminate among clinical stages of PCa progression.

To exclude any potential bias due to the different expression of DIhKPs in the three cell lines, we verified the differential abundance calculated in our previous work [19] using total LNCaP, PDB, and MDB extracts by MS. For all proteins, the concordance between differential expression and altered hnRNP K binding capacity (respectively, differentially expressed protein—DEP—and DIhKP columns in Table S4) was evaluated, excluding 16 proteins (highlighted in light grey in Table S4) from further analysis. Moreover, referring to the expression determined by immunohistochemistry (IHC) reported in "The Human Protein Atlas" (http://www.proteinatlas.org/), we verified that only three proteins (highlighted in heavy grey in Table S4) out of the 51 DIhKPs were not expressed in normal and tumor prostate tissues. According to these findings, 32 DIhKPs were retained and considered for further evaluations.

By analyzing AR-related profiles reported in the literature evaluating AR signature [36], androgen-responsive genes [37], AR-controlled genes, and co-regulators [9,38], we verified that six out of the 32 DIhKPs retained for our analysis correlated with AR activity (Table S5).

Next, we evaluated differential expression, both at protein and transcript level, in human prostate tissues for the selected 32 DIhKPs (Table S5). Using gene expression data from four published datasets (expression profiling by array) available on Gene Expression Omnibus (GEO) portal (GSE32269, GSE101607, GSE29650, GSE66532), we performed the differential expression analyses to identify transcript alterations in tissues at different stages of PCa development. Interestingly, 53% of the 32 DIhKPs selected in our in vitro models were differentially expressed also in the first two datasets comparing non-tumoral tissue, PCa, and two subgroups of CRPC (AR-driven and non-AR-driven) bone metastases. By analyzing recent and exhaustive literature of proteomic studies in PCa tissues reported by Larkin et al. [39], we observed different expression of seven DIhKPs in PCa compared with non-tumoral or hyperplastic prostate tissues.

Combining all data reported in Table S5, we defined a hnRNP K-AR-related signature comprising eight DIhKPs selected on the basis of their AR association as coregulators and/or interactors and their differential expression in both mCRPC subgroups and PCa (Table 2).

Table 2. Selected DIhKPs associated with hK-TF, AR, PCa and differential expression in human prostate tissues, delineating a hnRNP K–AR-related signature.

Gene symbol/Protein Name	Differential hnRNP K Interaction	Associated hK-TF	Association with AR	Human Prostate Tissues Differential Expression				Association with PCa (Literature Evidences)
				Proteomic Studies	Transcriptomic Studies			
KLK3/PSA	MDB&PDB vs. LNCaP ↓ *	PAX5	AR controlled	PCa vs. BPH ↓	PCa & mCRPC vs. NT ↑	mCRPC vs. PCa ↓	mCRPC non-AR driven vs. AR driven ↓	Biomarker [40–42]
SORD/Sorbitol dehydrogenase	MDB&PDB vs. LNCaP ↓	PPARG	AR controlled		PCa vs. NT ↑	mCRPC vs. PCa ↓	mCRPC non-AR driven vs. AR driven ↓	Increased expression in high Gleason PCa; reduced expression after castration [43]
SPON2/Spondin2	MDB&PDB vs. LNCaP ↓	HOX13	AR controlled				mCRPC non-AR driven vs. AR driven ↓	Biomarker [44,45]
ATP1B1/Sodium-potassium-transporting ATPase subunit beta1	PDB vs. LNCaP ↑ **	PPARG, XBP1, PAX5, HOX13, CMYB	AR controlled					Frequently overexpressed and amplified in CRPC [46]
ACTN4/Actinin4	PDB vs. LNCaP ↓	PPARG, XBP1, PAX5, HOX13, ZIC2, CMYB	AR coactivator, direct interaction		PCa vs. NT ↓			Reduced expression in high-grade PCa nuclei [42,47,48]
HSPB1/HSP27	PDB vs. LNCaP ↓	XBP1, PAX5	AR coactivator, direct interaction	PCa vs. NT & BPH ↑	PCa vs. NT ↑			Role in PCa epithelial to mesenchymal transition and resistance [49–51]
KHDRBS1/Sam68	MDB vs. LNCaP ↓	PPARG, PAX5, ZIC2, CMYB	AR coactivator/corepressor					Role in regulating AR splice variant, cell proliferation, and survival to chemotherapeutic agents [52–54]
IMPDH2/Inosine-5-monophosphate dehydrogenase 2	MDB vs. LNCaP ↓	PPARG, XBP1, PAX5, HOX13, CMYB		PCa vs. NT ↑	PCa & mCRPC vs. NT ↑		PCa vs. mCRPC non-AR driven ↑	Role in PCa metastasis and progression [55,56]

↓ * downregulated; ↑ ** upregulated.

For all proteins, evidence in the literature, reported in Table 2, shows a clear association with PCa progression. KLK3, SORD, and SPON2 expression differentiates hormone-dependent PCa from the two mCRPC subgroups. IMPDH2 tissue expression can differentiate PCa and mCRPC non-AR driven subgroups. Although there was no evidence for differential expression in androgen-dependent PCa and mCRPC tissues, the other four proteins (ACTN4, ATP1B1, HSPB1, KHDRBS1) were associated with AR. Of note, three AR co-regulators showed decreased hnRNP K interaction with respect to the parental cell line, depending on the cellular context (ACTN4 and HSPB1 for PDB and KHDRBS1 for MDB cell line). All the hK-TFs are potentially involved in regulating the transcription of the eight selected proteins. Exclusively the expression of ACTN4, an AR coregulator, is regulated by all hK-TFs, while the other proteins are controlled by one or more hK-TF.

Globally, these observations suggest that the hnRNP K–AR-related signature identified in our in vitro models might be associated to advanced stages of PCa and could identify molecular mCRPC subgroups.

3. Discussion

In the last years, experimental evidence has demonstrated that hnRNP K overexpression and aberrant cytoplasmic localization have an important role in PCa progression [12]. Recent findings suggest a relationship between AR and hnRNP K, as the latter regulates AR expression [57,58] and mRNA translation [15,24] and shapes a complex with AR and DNA modulated by anti-androgens [20]. In PCa tissues, the association of AR with cytoplasmic hnRNP K has a potential prognostic value [14], while the deregulation of the AKT–hnRNPK–AR–β-catenin pathway is involved in neuroendocrine differentiation [17]. Despite these evidence, the role of hnRNP K in androgen resistance has not been established yet.

HnRNP K is a modular protein that regulates a wide range of biological processes, among which transcription, splicing, and translation involved in the regulation of gene expression. By its ability to interact with both nucleic acids and proteins, hnRNP K acts like a docking platform coordinating the cross-talk between cell signaling pathways [10,21]. Consequently, the hnRNP K interactome is dynamic and changes depending on cellular compartments and external stimuli [27]. Moreover, hnRNP K functional versatility is regulated by several PTMs that modify its binding with nucleic acid and proteins [11,26].

On the basis of these premises, we explored the role of hnRNP K in androgen-resistance acquisition using three PCa cell lines: the androgen-responsive LNCaP and two resistant cell lines representing in vitro CRPC models, namely, the androgen-hypersensitive PDB and the androgen-insensitive MDB cell lines [19].

In LNCaP cell line, hnRNP K was found to regulate both AR expression and AR transcriptional activity (Figure 1a,b), while these functions were compromised in resistant cell lines, maximally in the most aggressive MDB. The adaptive phenotype characterizing PDB and MDB cells [19] under less stressful conditions (BIC removal) was found to drive a temporary recovery of AR activity which appeared to depend on the cellular context (Figure 1d,e). In the AR-inactive MDB cell line, restored androgen levels temporary reactivated the AR axis, determining a PSA rise correlated with AR and hnRNP K overexpression. Conversely, in PDB cell line, BIC removal increased PSA expression independently of AR and hnRNP K but associated with increased hnRNP K phosphorylation (Figure 2b) that could drive the AR hypersensitivity acquired during prolonged BIC exposure in androgen-containing medium.

The functional flexibility of hnRNP K arises, as for all the others hnRNPs, from the expression of alternatively spliced isoforms, PTMs, and subcellular distribution, which modify its binding activity to nucleic acids and proteins [59]. In response to mitogenic stimuli, the phosphorylated isoforms 1 and 2 generated by hnRNP K alternative splicing were differentially distributed in the nucleus and the cytoplasm [26], whereas the isoform 2 was downregulated in the colon carcinoma cell line HCT116 after p53 activation [60]. In the present study, we demonstrate a role for hnRNP K in ADT resistance acquisition depending on its alternative splicing, phosphorylation, and subcellular localization (Table 1 and Figure S2). Regardless of the AR status, the resistant cell lines showed higher levels of nuclear isoforms 1 and cytoplasmic isoform 2 than the androgen-responsive LNCaP. A quantitative comparison of the nuclear profiles of the pS116, pS284, and pS353 isoforms in the androgen-responsive LNCaP cell lines and in the resistant cell lines with different AR status and androgen responsiveness (Figure 3 and Table 1) showed a relationship between hnRNP K phosphorylation and AR transcriptional activity. In particular, to an active AR corresponded a low percentage of non-phosphorylated hnRNP K nuclear isoforms and vice versa (24% in LNCaP versus 66% in PC3). Interestingly, Ser116-phosphorylated isoforms could discriminate between resistant cell lines with different AR activity (nuclear fraction) and between LNCaP and all resistant cell lines (cytoplasmic fraction). It has been shown that the nuclear–cytoplasmic trafficking of hnRNP K may depend on Ser284 and Ser353 phosphorylation, determining inhibition of translation [24], and that Ser353 was involved in hnRNP K transactivation activity [61]. In line with these observations, our analysis showed an altered subcellular distribution of pS284 and pS353 in all CRPC cell lines in comparison to LNCaP.

Overall, these results suggest that the decreased phosphorylation of hnRNP K isoform 2, harboring the pS353 residue, and the subcellular distribution of pS116 and pS284 could have a significant role in ADT resistance.

It is increasingly evident that the identification of a molecular signature for specific stages of PCa would be useful not only for the selection of an appropriate and more effective therapy, but also for understanding the molecular basis of ADT resistance with relevant therapeutic implications [62]. The complex genomic and transcriptomic landscape of PCa has been explored through molecular analyses leading to the evidence that AR is involved in the development of drug resistance [61]. The selection pressure exerted by ADT induces alterations both in AR transcriptional program and in androgen-responsive gene (ARG) expression [37], triggering different downstream signaling pathways in PCa tissues that promote resistance [30,63]. Among the mechanisms proposed for AR reactivation during ADT, coregulators can play a critical role in influencing AR transcriptional activity and can represent new potential therapeutic targets for CRPC [64–67]. Recently, it was reported that AR coregulators control ARGs associated with different cellular processes in a context-dependent manner [9]. The dysregulated expression of AR coregulators in androgen-dependent PCa and in CRPC was reported to correlate with a poor prognosis and a more aggressive disease [65]. Therefore, it is reasonable to assume that the altered hnRNP K expression and function, by modifying the interaction with its partners involved in protein expression regulation (Figure 4 and Table S3), could alter the AR transcriptional program and androgen responsiveness in all stages of PCa.

In addition to AR coregulators, Obinata et al. [68] suggested a role for TFs in the control of the altered AR transcriptional program during anti-androgen treatment. The transcriptional regulatory networks of hK-TFs and DIhKPs subsets showed in Figure 5 confirm this hypothesis, since the composition of the hnRNP K interactome changed dynamically from LNCaP to the resistant cell lines PDB (AR-hypersensitive) and MDB (AR-inactive). Among the six hK-TFs, AR-controlled PPARG and HOX13 can be associated to ADT resistance. A recent study [35] has reported that PPARG controls prostate cell growth and differentiation and is involved in androgen-dependent PCa and CRPC evolution, acting as a tumor suppressor through its interplay with AR. The TF HOXA13, whose expression is associated with unfavorable survival [32], and the HOXB13, which controls PCa cell growth by inhibiting AR signaling [31], are also involved in PCa evolution.

The findings that emerge from our study demonstrate, for the first time, a clear-cut relationship between hnRNP K and AR, since, among the DIhKPs and the six hK-TFs, we found three AR coregulators (ACTN4, HSB1, KHDRBS1), four AR-controlled proteins (KLK3, SORD, SPON2, ATP1B1), and three TFs associated with AR activity (PPARG, ZIC2, HOX13) (Figure 6 and Table 2). In both androgen-resistant cell lines, the main hK-TF was PPARG which controls two distinct DIhKPs sets: in PDB (Figure 5a), the expression of eight upregulated DIhKPs was associated with the downregulation of the AR coregulators ACTN4 and HSB1, while, in MDB cell line (Figure 5b), all proteins under PPARG control were downregulated, except for PDIA4, including the AR coregulator KHDRBS1. Thus, we can infer that hnRNP K acts as a coordinator between AR and PPARG pathways and, depending on the AR coregulator involved, differently regulates the AR axis in androgen-dependent or -independent cell lines. As a confirmation, Moss et al. [69] reported different AR activity regulation by the PPARG ligand ciglitazone depending on androgen sensitivity, while Yang et al. [70] described that hnRNP K and KHDRBS1 interaction can regulate their activities in signal transduction pathways.

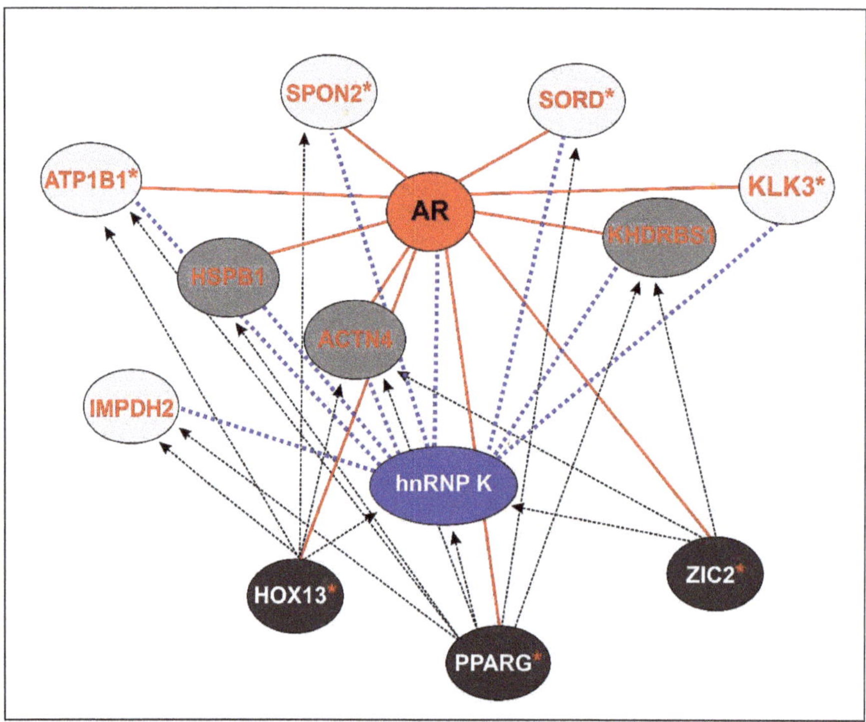

Figure 6. Schematic representation of the cross-talk network between hnRNP K, AR, the three AR-regulated hK-TFs, and the eight proteins selected for the hnRNP K–AR-related signature. The eight DIhKPs (heavy and light gray circles) directly or indirectly interact with hnRNP K (blue dotted lines). The hK-TFs (black circles) potentially control hnRNP K, and the eight DIhKPs (black dotted lines). AR coregulators (heavy gray circles) and androgen-responsive gene (ARGs) (highlighted with red *) are associated with AR (red solid lines).

Understanding the context and the functional role of key proteins in ADT resistance could allow the identification of a molecular signature, rather than single proteins, that could overcome the main limit of prostate cancer treatments: tumor heterogeneity. Previously, we demonstrated that, in our CRPC model, the AR transcriptional program was modified by ADT [19]. In this study, we found that hnRNP K expression, alternatively spliced isoforms localization, and impaired phosphorylation promote ADT resistance depending on the cellular context and external stimuli. Using different analytical strategies, we identified eight proteins with dissimilar roles in the AR axis and correlation with clinical PCa progression. Of note, KLK3, SORD, and SPON2 can discriminate between non-AR-driven and AR-driven mCRPC and are characterized by common features: (i) decreased interaction with hnRNP K; (ii) control by AR; (iii) association with a single hK-TF (PAX5, PPARG, HOX13).

The eight proteins highlighted in this study (KLK3, SORD, SPON2, IMPDH2, ACTN4, ATP1B1, HSPB1, and KHDRBS1) represent a signature associated with AR modulation during the transition from androgen-dependent to castration-resistant prostate cancers. These proteins might discriminate specific stages of PCa and thus predict patients' prognosis and response to treatments. Further investigations on PCa tissues or liquid biopsies from CRPC patients are needed to validate the clinical usefulness of the hnRNP K–AR-related signature and to evaluate whether some of the proteins included in our signature might become potential targets for new drugs development.

4. Materials and Methods

4.1. Cell Culture

The human prostate cancer cell lines LNCaP, 22Rv1, and PC3 were purchased from the American Type Culture Collection (CRL-1740, CRL-2505, CRL1435) and cultured in RPMI (Celbio, Milan, Italy) medium containing 10% fetal bovine serum, 2 mM glutamine, 1% penicillin, and 1% streptomycin. The LNCaP and 22Rv1 medium was also supplemented with 10 mM HEPES, 10 mM sodium pyruvate, and 0.25% glucose. The passage numbers at which LNCaP cells were used ranged from 22 to 30. The BIC-resistant cell lines PDB and MDB, obtained in our laboratory [19], were cultured in RPMI medium without phenol red, containing 10% charcoal-stripped fetal bovine serum, 2 mM glutamine, 10 mM HEPES, 10 mM sodium pyruvate, 0.25% glucose, 1% penicillin, 1% streptomycin, and 10 μM BIC. The PDB cell line was supplemented with 0.1 nM DHT.

4.2. Cell Fractionation

All buffers were supplemented with a protease inhibitors cocktail (5 mM $Na_2S_2O_5$, 1 mM phenylmethylsulfonyl fluoride, 0.5 mM benzamidine, 20 mg/mL leupeptin, 10 mg/mL pepstatin A, 25 mg/mL aprotinin), 1 mM Na_3VO_4, and 1 mM dithiothreitol. The fractionation procedures were performed at 4 °C. The cells were mechanically harvested with a sterile cell scraper and washed three times in 30 mL of PBS. Total protein extracts, obtained using RIPA buffer, and cytoplasmic and nuclear extracts were prepared as reported [16,20]. After cell fractionation, the pellet was delipidated, cleaned, and solubilized for 1D or 2D gel electrophoresis as previously described [20]. The protein concentrations were determined using the Bio-Rad (Hercules, CA, USA) protein microassay with bovine serum albumin as a standard.

4.3. 1D and 2D Gel Electrophoresis and WB Analyses

Eight micrograms of proteins extracted from different cell fractions were loaded onto an 8–14% linear gradient (1D) or 7.5% (1D Phos-tag) polyacrylamide gels, containing 25 mM polyacrylamide-bound Mn^{2+}-Phos-tag^{TM} ligand (Phos-tag Consortium, Hiroshima, Japan) and separated at 5 mA/gel for 16 h at a constant temperature of 12 °C. 2D Phos-tag was performed as described [20]. After electrophoresis separation, the proteins were transferred to a Hybond-P membrane (GE Healthcare, Piscataway, NJ, USA) and probed with anti-hnRNP K (Santa Cruz Biotechnology Inc., Santa Cruz, CA, USA; 1:5000), anti-AR (DAKO, Carpinteria, CA, USA; 1:500), and anti-PSA (Cell Signaling Technology, Danvers, MA, USA; 1:2000) antibodies at 4 °C overnight. HRP-conjugated secondary antibodies (Cell Signaling Technologies, Danvers, MA, USA; 1:2000) were used, and protein bands were detected by chemiluminescent HRP substrate (Immobilon Western, Millipore, Darmstadt, Germany) and Hyper film-ECL (GE-healthcare, Lafayette, CO, USA), according to the manufacturer's instructions. As PDB and MDB cell lines showed a global downregulation of protein expression compared to LNCaP cell line [19], the conventional normalization approach was inadequate. Different results were in fact obtained using β-actin- or Sypro Ruby- (Invitrogen, Eugene, OR, USA) normalized data to evaluate the relative amounts of AR and hnRNP K in total extracts from LNCaP, PDB, and MDB (Figure S5). For 1D analysis, the relative amount of immunoreactive bands on WB films scanned with a GS-800 densitometer (BioRad, Hercules, CA, USA), was assessed using Bio-Rad Quantity One software and normalizing the optical density of each band to the total optical density of the corresponding lane in the Sypro Ruby-stained gel providing the total protein content [16]. In Figure S6, the Sypro Ruby-stained gels used for quantitative analysis of WB reported in Figure 1a,d and Figure 2a are shown. The quantitative analysis of 1D Phos-tag WB was carried out comparing the relative percentage of phK0, phK1, and phK2 hnRNP K isoforms in each cell lines. The 2D Phos-tag analyses, including detection, alignment, and matching for spots present in all 2D maps, were done using the software package PDQuest (ver. 8.6, BioRad, Hercules, CA, USA).

Statistical significance was evaluated by two-tailed Student's *t*-test within OriginPro 7.5 software (OriginLab, Northampton, MA, USA), and the differences were considered statistically significant when $p < 0.05$.

4.4. hnRNP K Silencing and Reporter Assay

ON-TARGET plus SMART pool for human hnRNP K (Dharmacon, GE- healthcare, Lafayette, CO, USA) was used to knockdown the expression of hnRNP K; siCONTROL non-targeting (NT) siRNA pool (Dharmacon, Lafayette, CO, USA) was used as a negative control. The cells were grown in their proper medium and transfected with Lipofectamine 2000 (Thermo Fisher, Waltham, MA, USA) following the manufacturer's instructions. Protein expression was analyzed by WB 72 h after transfection.

To measure AR function, the cells were transfected with Lipofectamine 2000 utilizing the Cignal androgen receptor dual-luciferase reporter kit (Qiagen, Hilden, Germany), following the manufacturer's instructions. Luciferase activity was assayed in triplicate 48 h after transfection using the dual-luciferase reporter assay kit (Promega, Madison, WI, USA). To control for general effects on transcription, Renilla luciferase was co-transfected in all reporter assays, and luciferase values represent the ratio luciferase/Renilla.

4.5. Co-Immunoprecipitation

Co-immunoprecipitation, employed to identify proteins directly or indirectly interacting with hnRNP K, was carried out using the Pierce Direct IP kit according to the manufacturer's instructions (Pierce Biotechnology, Rockford, IL, USA). Briefly, total cellular extracts, obtained with IP Lysis/Wash buffer containing a protease inhibitor cocktail, were pre-cleared with agarose resin and passed through an AminoLink plus resin with covalently immobilized anti-hnRNP K antibodies (Santa Cruz Biotechnology, Santa Cruz, CA, USA). The eluted proteins were subjected to MS for identification of the hnRNP K interactome. Control resin and quenched antibody-coupling resin without anti-hnRNP K antibodies were utilized to identify non-specific interactions. Common proteins between these groups and the co-IPs were excluded from bioinformatic analyses. For MS analysis, three independent experiments with LNCaP, PDB, and MDB were performed.

The hnRNP K co-IPs were analyzed by WB for hnRNP K and AR expression evaluation in LNCaP, PDB, and MDB cell lines (Figure S3).

4.6. MS Analysis

The samples were processed by the FASP Protein Digestion Kit (Expedeon, Harston, UK). Briefly, the co-IP samples were mixed with 0.3 mL of 8 M urea in 0.1 M Tris/HCl pH 8.5 (UA solution), loaded into the filtration devices, and alkylated in 0.1 mL of 50 mM iodoacetamide in UA solution for 1 h in darkness at room temperature. The samples were digested using, sequentially, 1 µg of LysC and 3 µg of Trypsin in 50 mM NaHCO$_3$ solution at 37 °C overnight. Peptides were collected by centrifugation of the filter units for 10 min, and the filter devices were rinsed with two 40 µL washes of 50 mM NaHCO$_3$ and 50 µL 0.5 M NaCl to eliminate the hydrophobic interactions. Each sample digest was desalted on StageTips and analyzed by liquid chromatography–tandem mass spectrometry (LC–MS/MS).

All MS experiments were performed on a nanoscale high-performance liquid chromatography system connected to a hybrid linear trap quadrupole (LTQ) Orbitrap mass spectrometer. The MS instrument was operated in data-dependent mode to automatically switch between full-scan MS and MS/MS acquisition. Survey full-scan MS spectra were acquired in the Orbitrap analyzer with resolution $R = 60,000$. The 10 most intense peptide ions with charge states ≥ 2 were sequentially isolated and fragmented by collision-induced dissociation in the LTQ mass spectrometer. The raw mass spectrometric data were analyzed using the MaxQuant pipeline. MaxQuant enables high peptide identification rates, individualized p.p.b.-range mass accuracies, and proteome-wide protein quantification [71]. The statistical and the pathways analyses were done with the freely available Perseus [72] and Cytoscape software [73] respectively.

4.7. Network Analysis

Cytoscape (v 3.6.0) was used to visualize the protein–protein interaction network obtained from the above proteomic data (co-IP) and to highlight functional annotations [28]. Statistically significant proteins were used as the queries to interrogate Intact and Reactome database in order to obtain a protein–protein interaction network. The network was treated as undirected, and all duplicated edges and self-loop were removed. To highlight the most important proteins, a network analysis was performed. In particular, the degrees and closeness were assessed. The degree is the simplest topological index, corresponding to the number of nodes adjacent to a given node, where "adjacent" means directly connected. The degree allows an immediate evaluation of the regulatory relevance of the node. The closeness is a node centrality index. The closeness of a specific node is calculated by computing the shortest path between the node and all other nodes in the graph and then calculating the summa. From the biological point of view, the closeness can be interpreted as the "probability" of a protein to be functionally relevant for several other proteins. Moreover, the MaxLFQ expression data were loaded into the network as attribute node. Subsequently, a clusterization of significant proteins related to the expression data and to the topology of the network was performed as previously described, by clustermaker2 (Cytoscape App) [28] with the "autoSOME Clustering" algorithm (node attribute = expression; distance metric: Euclidean, ignore node with no data). Each obtained cluster was functionally tagged by an enrichment, performed by JEPETTO [29] (network: PSICQUIC, reference database: GO MF/prostate tissue) if statistically significant results were obtained.

4.8. Differential Expression and Enrichment Analyses

Differential expression analyses were carried out both on protein detected through MS spectroscopy for their binding to hnRNP K and on gene expression data extracted from four datasets published and available on the GEO portal (GSE32269, GSE101607, GSE29650, GSE66532).

In the first case, the log-transformed values of protein binding were imported and analyzed through linear model fitting and empirical Bayes approach provided by Limma R package [74]. Three different contrast matrices were generated: PDB versus LNCaP, MDB versus LNCaP, MDB versus PDB. Proteins showing a p-value ≤ 0.05 and a $|\log_2 FC| \geq 1$ were considered differentially interacting (DIhKPs).

Microarray data from GEO datasets were imported through GEOquery R package [75], "GEOquery: a bridge between the Gene Expression Omnibus (GEO) and BioConductor", and comparison analysis of genes was performed using Limma.

For each dataset, pairwise comparisons between sample groups, as indicated in Table S5, were performed. Genes having a $|\log_2 FC| \geq 1$ and the corresponding p-value ≤ 0.05 were considered statistically significant.

Enrichment analysis was performed on the 51 DIhKPs to identify the over-represented GO annotations regarding Biological Processes and Molecular Function, as well as the transcription factor binding sites through the category UCSC-TFBS provided by the functional annotation tool DAVID v 6.8 [76].

Supplementary Materials: Supplementary materials can be found at http://www.mdpi.com/1422-0067/19/7/1920/s1.

Author Contributions: Conceptualization, P.B.; Data curation, M.C. and I.G.; Formal analysis, I.G. and E.I.; Funding acquisition, M.R.G., A.P., N.F., F.B., and P.B.; Investigation, M.C., I.G., E.I., N.F., and P.B.; Project administration, P.B.; Resources, M.C. and N.F.; Supervision, M.R.G., A.P., and P.B.; Writing–original draft, P.B.; Writing–review & editing, C.C., N.F., and F.B.

Funding: This work was supported by Compagnia di San Paolo (2012.1588), IRCCS AOU San Martino-IST (Project 5x1000), MIUR PON02-00619, and National Research University Higher School of Economics RSF Grant 14-41-00039.

Conflicts of Interest: The authors declare no conflict of interest.

Abbreviations

ADT	Androgen deprivation therapy
AR	Androgen receptor
ARG	Androgen-responsive gene
BIC	Bicalutamide
Co-IP	Co-immunoprecipitate
CRPC	Castration-resistant prostate cancer
DAVID	Database for Annotation Visualization and Integrated Discovery
DEP	Differentially expressed protein
DIhKP	Differentially interacting hnRNP K protein
DHT	5-α-dihydrotestosterone
GO	Gene ontology
hK-TF	Transcription factor potentially regulating the expression of differentially interacting hnRNP K protein
LTQ	Linear trap quadrupole
MS	Mass spectrometry
PTM	Post-translation modifications
TF	Transcription factor
TFBS	Transcription factor binding sites
PAGE	Polyacrylamide gel electrophoresis
PCa	Prostate cancer
WB	Western blotting

References

1. Katsogiannou, M.; Ziouziou, H.; Karaki, S.; Andrieu, C.; Henry de Villeneuve, M.; Rocchi, P. The hallmarks of castration-resistant prostate cancers. *Cancer Treat. Rev.* **2015**, *41*, 588–597. [CrossRef] [PubMed]

2. Cattrini, C.; Zanardi, E.; Vallome, G.; Cavo, A.; Cerbone, L.; Di Meglio, A.; Fabbroni, C.; Latocca, M.M.; Rizzo, F.; Messina, C.; et al. Targeting androgen-independent pathways: New chances for patients with prostate cancer? *Crit. Rev. Oncol. Hematol.* **2017**, *118*, 42–53. [CrossRef] [PubMed]

3. Zarif, J.C.; Miranti, C.K. The importance of non-nuclear AR signaling in prostate cancer progression and therapeutic resistance. *Cell Signal.* **2016**, *28*, 348–356. [CrossRef] [PubMed]

4. Shah, R.B.; Mehra, R.; Chinnaiyan, A.M.; Shen, R.; Ghosh, D.; Zhou, M.; Macvicar, G.R.; Varambally, S.; Harwood, J.; Bismar, T.A.; et al. Androgen-independent prostate cancer is a heterogeneous group of diseases: Lessons from a rapid autopsy program. *Cancer Res.* **2004**, *64*, 9209–9216. [CrossRef] [PubMed]

5. Beltran, H.; Yelensky, R.; Frampton, G.M.; Park, K.; Downing, S.R.; MacDonald, T.Y.; Jarosz, M.; Lipson, D.; Tagawa, S.T.; Nanus, D.M.; et al. Targeted next-generation sequencing of advanced prostate cancer identifies potential therapeutic targets and disease heterogeneity. *Eur. Urol.* **2012**, *63*, 920–926. [CrossRef] [PubMed]

6. Ciccarese, C.; Massari, F.; Iacovelli, R.; Fiorentino, M.; Montironi, R.; Di Nunno, V.; Giunchi, F.; Brunelli, M.; Tortora, G. Prostate cancer heterogeneity: Discovering novel molecular targets for therapy. *Cancer Treat. Rev.* **2017**, *54*, 68–73. [CrossRef] [PubMed]

7. Wyatt, A.W.; Gleave, M.E. Targeting the adaptive molecular landscape of castration-resistant prostate cancer. *EMBO Mol. Med.* **2015**, *7*, 878–894. [CrossRef] [PubMed]

8. Emmons, M.F.; Faiao-Flores, F.; Smalley, K.S. The role of phenotypic plasticity in the escape of cancer cells from targeted therapy. *Biochem. Pharmacol.* **2016**, *122*, 1–9. [CrossRef] [PubMed]

9. Liu, S.; Kumari, S.; Hu, Q.; Senapati, D.; Venkadakrishnan, V.B.; Wang, D.; DePriest, A.D.; Schlanger, S.E.; Ben-Salem, S.; Valenzuela, M.M.; et al. A comprehensive analysis of coregulator recruitment, androgen receptor function and gene expression in prostate cancer. *Elife* **2017**, *6*, e28482. [CrossRef] [PubMed]

10. Bomsztyk, K.; Denisenko, O.; Ostrowski, J. hnRNP K: One protein multiple processes. *Bioessays* **2004**, *26*, 629–638. [CrossRef] [PubMed]

11. Mikula, M.; Ostrowski, J. Heterogeneous Nuclear Ribonucleoprotein K—Hnrnp K. The Transcription Factor Encyclopedia. *Genome Biol.* **2012**, *13*, R24. Available online: http://www.cisreg.ca/cgi-bin/tfe/articles.pl?tfid=551 (accessed on 27 November 2017).

12. Barboro, P.; Ferrari, N.; Balbi, C. Emerging roles of heterogeneous nuclear ribonucleoprotein K (hnRNP K) in cancer progression. *Cancer Lett.* **2014**, *352*, 152–159. [CrossRef] [PubMed]

13. Barboro, P.; Repaci, E.; Rubagotti, A.; Salvi, S.; Boccardo, S.; Spina, B.; Truini, M.; Introini, C.; Puppo, P.; Ferrari, N.; et al. Heterogeneous nuclear ribonucleoprotein K: Altered pattern of expression associated with diagnosis and prognosis of prostate cancer. *Br. J. Cancer* **2009**, *100*, 1608–1616. [CrossRef] [PubMed]

14. Barboro, P.; Salvi, S.; Rubagotti, A.; Boccardo, S.; Spina, B.; Truini, M.; Carmignani, G.; Introini, C.; Ferrari, N.; Boccardo, F.; et al. Prostate cancer: Prognostic significance of the association of heterogeneous nuclear ribonucleoprotein K and androgen receptor expression. *Int. J. Oncol.* **2014**, *44*, 1589–1598. [CrossRef] [PubMed]

15. Mukhopadhyay, N.K.; Kim, J.; Cinar, B.; Ramachandran, A.; Hager, M.H.; Di Vizio, D.; Adam, R.M.; Rubin, M.A.; Raychaudhuri, P.; De Benedetti, A.; et al. Heterogeneous nuclear ribonucleoprotein K is a novel regulator of androgen receptor translation. *Cancer Res.* **2009**, *69*, 2210–2218. [CrossRef] [PubMed]

16. Barboro, P.; Repaci, E.; Ferrari, N.; Rubagotti, A.; Boccardo, F.; Balbi, C. Androgen receptor and heterogeneous nuclear ribonucleoprotein K colocalize in the nucleoplasm and are modulated by bicalutamide and 4-hydroxy-tamoxifen in prostatic cancer cell lines. *Prostate* **2011**, *71*, 1466–1479. [CrossRef] [PubMed]

17. Ciarlo, M.; Benelli, R.; Barbieri, O.; Minghelli, S.; Barboro, P.; Balbi, C.; Ferrari, N. Regulation of neuroendocrine differentiation by AKT/hnRNPK/AR/beta-catenin signaling in prostate cancer cells. *Int. J. Cancer* **2012**, *131*, 582–590. [CrossRef] [PubMed]

18. Kelly, K.; Balk, S.P. Reprogramming to resist. *Science* **2017**, *355*, 29–30. [CrossRef] [PubMed]

19. Ferrari, N.; Granata, I.; Capaia, M.; Piccirillo, M.; Guarracino, M.R.; Vene, R.; Brizzolara, A.; Petretto, A.; Inglese, E.; Morini, M.; et al. Adaptive phenotype drives resistance to androgen deprivation therapy in prostate cancer. *Cell Commun. Signal.* **2017**, *15*, 51. [CrossRef] [PubMed]

20. Barboro, P.; Borzi, L.; Repaci, E.; Ferrari, N.; Balbi, C. Androgen receptor activity is affected by both nuclear matrix localization and the phosphorylation status of the heterogeneous nuclear ribonucleoprotein K in anti-androgen-treated LNCaP cells. *PLoS ONE* **2013**, *8*, e79212. [CrossRef] [PubMed]

21. Moumen, A.; Magill, C.; Dry, K.L.; Jackson, S.P. ATM-dependent phosphorylation of heterogeneous nuclear ribonucleoprotein K promotes p53 transcriptional activation in response to DNA damage. *Cell Cycle* **2013**, *12*, 698–704. [CrossRef] [PubMed]

22. Habelhah, H.; Shah, K.; Huang, L.; Burlingame, A.L.; Shokat, K.M.; Ronai, Z. Identification of new JNK substrate using ATP pocket mutant JNK and a corresponding ATP analogue. *J. Biol. Chem.* **2001**, *276*, 18090–18095. [CrossRef] [PubMed]

23. Laury-Kleintop, L.D.; Tresini, M. Hammond, O. Compartmentalization of hnRNP-K during cell cycle progression and its interaction with calponin in the cytoplasm. *J. Cell. Biochem.* **2005**, *95*, 1042–1056. [CrossRef] [PubMed]

24. Habelhah, H.; Shah, K.; Huang, L.; Ostareck-Lederer, A.; Burlingame, A.L.; Shokat, K.M.; Hentze, M.W.; Ronai, Z. ERK phosphorylation drives cytoplasmic accumulation of hnRNP-K and inhibition of mRNA translation. *Nat. Cell Biol.* **2001**, *3*, 325–330. [CrossRef] [PubMed]

25. Sampson, N.; Neuwirt, H.; Puhr, M.; Klocker, H.; Eder, I.E. In vitro model systems to study androgen receptor signaling in prostate cancer. *Endocr. Relat. Cancer* **2013**, *20*, R49–R64. [CrossRef] [PubMed]

26. Kimura, Y.; Nagata, K.; Suzuki, N.; Yokoyama, R.; Yamanaka, Y.; Kitamura, H.; Hirano, H.; Ohara, O. Characterization of multiple alternative forms of heterogeneous nuclear ribonucleoprotein K by phosphate-affinity electrophoresis. *Proteomics* **2010**, *10*, 3884–3895. [CrossRef] [PubMed]

27. Mikula, M.; Dzwonek, A.; Karczmarski, J.; Rubel, T.; Dadlez, M.; Wyrwicz, L.S.; Bomsztyk, K.; Ostrowski, J. Landscape of the hnRNP K protein-protein interactome. *Proteomics* **2006**, *6*, 2395–2406. [CrossRef] [PubMed]

28. Shannon, P.; Markiel, A.; Ozier, O.; Baliga, N.S.; Wang, J.T.; Ramage, D.; Amin, N.; Schwikowski, B.; Ideker, T. Cytoscape: A software environment for integrated models of biomolecular interaction networks. *Genome Res.* **2003**, *13*, 2498–2504. [CrossRef] [PubMed]

29. Winterhalter, C.; Widera, P.; Krasnogor, N. JEPETTO: A Cytoscape plugin for gene set enrichment and topological analysis based on interaction networks. *Bioinformatics* **2013**, *30*, 1029–1030. [CrossRef] [PubMed]

30. Sharma, N.L.; Massie, C.E.; Ramos-Montoya, A.; Zecchini, V.; Scott, H.E.; Lamb, A.D.; MacArthur, S.; Stark, R.; Warren, A.Y.; Mills, I.G.; et al. The androgen receptor induces a distinct transcriptional program in castration-resistant prostate cancer in man. *Cancer Cell* **2013**, *23*, 35–47. [CrossRef] [PubMed]

31. Kim, Y.R.; Oh, K.J.; Park, R.Y.; Xuan, N.T.; Kang, T.W.; Kwon, D.D.; Choi, C.; Kim, M.S.; Nam, K.I.; Ahn, K.Y.; et al. HOXB13 promotes androgen independent growth of LNCaP prostate cancer cells by the activation of E2F signaling. *Mol. Cancer* **2010**, *9*, 124. [CrossRef] [PubMed]

32. Dong, Y.; Cai, Y.; Liu, B.; Jiao, X.; Li, Z.T.; Guo, D.Y.; Li, X.W.; Wang, Y.J.; Yang, D.K. HOXA13 is associated with unfavorable survival and acts as a novel oncogene in prostate carcinoma. *Future Oncol.* **2017**, *13*, 1505–1516. [CrossRef] [PubMed]

33. Li, L.; Chang, W.; Yang, G.; Ren, C.; Park, S.; Karantanos, T.; Karanika, S.; Wang, J.; Yin, J.; Shah, P.K.; et al. Targeting poly(ADP-ribose) polymerase and the c-Myb-regulated DNA damage response pathway in castration-resistant prostate cancer. *Sci. Signal.* **2015**, *7*, ra47. [CrossRef] [PubMed]

34. Alghadi, A.Y.; Khalil, A.M.; Alazab, R.S.; Aldaoud, N.H.; Zyoud, A.M.-S. Unique Expression of the XBP1 Gene Correlates with with Human Prostate Cancer. *J. Investig. Genom.* **2017**, *4*, 00091. [CrossRef]

35. Olokpa, E.; Moss, P.E.; Stewart, L.V. Crosstalk between the Androgen Receptor and PPAR Gamma Signaling Pathways in the Prostate. *PPAR Res.* **2017**, *2017*, 9456020. [CrossRef] [PubMed]

36. Mendiratta, P.; Mostaghel, E.; Guinney, J.; Tewari, A.K.; Porrello, A.; Barry, W.T.; Nelson, P.S.; Febbo, P.G. Genomic strategy for targeting therapy in castration-resistant prostate cancer. *J. Clin. Oncol.* **2009**, *27*, 2022–2029. [CrossRef] [PubMed]

37. Tien, A.H.; Saladar, M.D. *Androgen-Responsive Gene Expression in Prostate Cancer Progression*, 1st ed.; Springer: New York, NY, USA, 2013; pp. 135–153. ISBN 978-1-4614-6182-1.

38. Beitel, L. Androgen Receptor Interacting Proteins and Coregulators Table. 2010. Available online: http://androgendb.mcgill.ca/ARinteract.pdf (accessed on 5 December 2017).

39. Larkin, S.E.; Zeidan, B.; Taylor, M.G.; Bickers, B.; Al-Ruwaili, J.; Aukim-Hastie, C.; Townsend, P.A. Proteomics in prostate cancer biomarker discovery. *Expert Rev. Proteom.* **2018**, *7*, 93–102. [CrossRef] [PubMed]

40. Prensner, J.R.; Rubin, M.A.; Wei, J.T.; Chinnaiyan, A.M. Beyond PSA: The next generation of prostate cancer biomarkers. *Sci. Transl. Med.* **2012**, *4*, 127rv3. [CrossRef] [PubMed]

41. Hatakeyama, S.; Yoneyama, T.; Tobisawa, Y.; Ohyama, C. Recent progress and perspectives on prostate cancer biomarkers. *Int. J. Clin. Oncol.* **2017**, *22*, 214–221. [CrossRef] [PubMed]

42. Intasqui, P.; Bertolla, R.P.; Sadi, M.V. Prostate cancer proteomics: Clinically useful protein biomarkers and future perspectives. *Expert Rev. Proteom.* **2017**, *15*, 65–79. [CrossRef] [PubMed]

43. Szabo, Z.; Hamalainen, J.; Loikkanen, I.; Moilanen, A.M.; Hirvikoski, P.; Vaisanen, T.; Paavonen, T.K.; Vaarala, M.H. Sorbitol dehydrogenase expression is regulated by androgens in the human prostate. *Oncol. Rep.* **2010**, *23*, 1233–1239. [CrossRef] [PubMed]

44. Qian, X.; Li, C.; Pang, B.; Xue, M.; Wang, J.; Zhou, J. Spondin-2 (SPON2), a more prostate-cancer-specific diagnostic biomarker. *PLoS ONE* **2012**, *7*, e37225. [CrossRef] [PubMed]

45. Zhu, B.P.; Guo, Z.Q.; Lin, L.; Liu, Q. Serum BSP, PSADT, and Spondin-2 levels in prostate cancer and the diagnostic significance of their ROC curves in bone metastasis. *Eur. Rev. Med. Pharmacol. Sci.* **2017**, *21*, 61–67. [PubMed]

46. Vainio, P.; Wolf, M.; Edgren, H.; He, T.; Kohonen, P.; Mpindi, J.P.; Smit, F.; Verhaegh, G.; Schalken, J.; Perala, M.; et al. Integrative genomic, transcriptomic, and RNAi analysis indicates a potential oncogenic role for FAM110B in castration-resistant prostate cancer. *Prostate* **2011**, *72*, 789–802. [CrossRef] [PubMed]

47. Honda, K. The biological role of actinin-4 (ACTN4) in malignant phenotypes of cancer. *Cell Biosci.* **2015**, *5*, 41. [CrossRef] [PubMed]

48. Jasavala, R.; Martinez, H.; Thumar, J.; Andaya, A.; Gingras, A.C.; Eng, J.K.; Aebersold, R.; Han, D.K.; Wright, M.E. Identification of putative androgen receptor interaction protein modules: Cytoskeleton and endosomes modulate androgen receptor signaling in prostate cancer cells. *Mol. Cell Proteom.* **2007**, *6*, 252–271. [CrossRef] [PubMed]

49. Cordonnier, T.; Bishop, J.L.; Shiota, M.; Nip, K.M.; Thaper, D.; Vahid, S.; Heroux, D.; Gleave, M.; Zoubeidi, A. Hsp27 regulates EGF/beta-catenin mediated epithelial to mesenchymal transition in prostate cancer. *Int. J. Cancer* **2015**, *136*, E496–E507. [CrossRef] [PubMed]

50. Zoubeidi, A.; Zardan, A.; Beraldi, E.; Fazli, L.; Sowery, R.; Rennie, P.; Nelson, C.; Gleave, M. Cooperative interactions between androgen receptor (AR) and heat-shock protein 27 facilitate AR transcriptional activity. *Cancer Res.* **2007**, *67*, 10455–10465. [CrossRef] [PubMed]

51. Rocchi, P.; Beraldi, E.; Ettinger, S.; Fazli, L.; Vessella, R.L.; Nelson, C.; Gleave, M. Increased Hsp27 after androgen ablation facilitates androgen-independent progression in prostate cancer via signal transducers and activators of transcription 3-mediated suppression of apoptosis. *Cancer Res.* **2005**, *65*, 11083–11093. [CrossRef] [PubMed]

52. Stockley, J.; Markert, E.; Zhou, Y.; Robson, C.N.; Elliott, D.J.; Lindberg, J.; Leung, H.Y.; Rajan, P. The RNA-binding protein Sam68 regulates expression and transcription function of the androgen receptor splice variant AR-V7. *Sci. Rep.* **2015**, *5*, 13426. [CrossRef] [PubMed]

53. Frisone, P.; Pradella, D.; Di Matteo, A.; Belloni, E.; Ghigna, C.; Paronetto, M.P. SAM68: Signal Transduction and RNA Metabolism in Human Cancer. *Biomed Res. Int.* **2015**, *2015*, 528954. [CrossRef] [PubMed]

54. Busa, R.; Paronetto, M.P.; Farini, D.; Pierantozzi, E.; Botti, F.; Angelini, D.F.; Attisani, F.; Vespasiani, G.; Sette, C. The RNA-binding protein Sam68 contributes to proliferation and survival of human prostate cancer cells. *Oncogene* **2007**, *26*, 4372–4382. [CrossRef] [PubMed]

55. Zhou, L.; Xia, D.; Zhu, J.; Chen, Y.; Chen, G.; Mo, R.; Zeng, Y.; Dai, Q.; He, H.; Liang, Y.; et al. Enhanced expression of IMPDH2 promotes metastasis and advanced tumor progression in patients with prostate cancer. *Clin. Transl. Oncol.* **2014**, *16*, 906–913. [CrossRef] [PubMed]

56. Albitar, M.; Ma, W.; Lund, L.; Albitar, F.; Diep, K.; Fritsche, H.A.; Shore, N. Predicting Prostate Biopsy Results Using a Panel of Plasma and Urine Biomarkers Combined in a Scoring System. *J. Cancer* **2015**, *7*, 297–303. [CrossRef] [PubMed]

57. Brizzolara, A.; Benelli, R.; Vene, R.; Barboro, P.; Poggi, A.; Tosetti, F.; Ferrari, N. The ErbB family and androgen receptor signaling are targets of Celecoxib in prostate cancer. *Cancer Lett.* **2017**, *400*, 9–17. [CrossRef] [PubMed]

58. Wang, L.G.; Johnson, E.M.; Kinoshita, Y.; Babb, J.S.; Buckley, M.T.; Liebes, L.F.; Melamed, J.; Liu, X.M.; Kurek, R.; Ossowski, L.; et al. Androgen receptor overexpression in prostate cancer linked to Pur alpha loss from a novel repressor complex. *Cancer Res.* **2008**, *68*, 2678–2688. [CrossRef] [PubMed]

59. Hang, S.P.; Tang, Y.H.; Smith, R. Functional diversity of hnRNPs: Past, present and perspectives. *Biochem. J.* **2010**, *430*, 379–392. [CrossRef]

60. Rahman-Roblick, R.; Roblick, U.J.; Hellman, U.; Conrotto, P.; Liu, T.; Becker, S.; Hirschberg, D.; Jörnvall, H.; Auer, G.; Wiman, K.G. p53 targets identified by protein expression profiling. *Proc. Natl. Acad. Sci. USA* **2007**, *104*, 5401–5406. [CrossRef] [PubMed]

61. Chan, J.Y.H.; Huang, S.M.; Liu, S.T.; Huang, C.H. The transactivation domain of heterogeneous nuclear ribonucleoprotein K overlaps its nuclear shuttling domain. *Int. J. Biochem. Cell Biol.* **2008**, *40*, 2078–2089. [CrossRef] [PubMed]

62. Vlachostergios, P.J.; Puca, L.; Beltran, H. Emerging Variants of Castration-Resistant Prostate Cancer. *Curr. Oncol. Rep.* **2017**, *19*, 32. [CrossRef] [PubMed]

63. Lamb, A.D.; Massie, C.E.; Neal, D.E. The transcriptional programme of the androgen receptor (AR) in prostate cancer. *BJU Int.* **2014**, *113*, 358–366. [CrossRef] [PubMed]

64. Holzbeierlein, J.; Lal, P.; LaTulippe, E.; Smith, A.; Satagopan, J.; Zhang, L.; Ryan, C.; Smith, S.; Scher, H.; Scardino, P.; et al. Gene expression analysis of human prostate carcinoma during hormonal therapy identifies androgen-responsive genes and mechanisms of therapy resistance. *Am. J. Pathol.* **2004**, *164*, 217–227. [CrossRef]

65. Coutinho, I.; Day, T.K.; Tilley, W.D.; Selth, L.A. Androgen receptor signaling in castration-resistant prostate cancer: A lesson in persistence. *Endocr. Relat. Cancer* **2016**, *23*, T179–T197. [CrossRef] [PubMed]

66. Shiota, M.; Yokomizo, A.; Fujimoto, N.; Naito, S. Androgen receptor cofactors in prostate cancer: Potential therapeutic targets of castration-resistant prostate cancer. *Curr. Cancer Drug Targets* **2011**, *11*, 870–881. [CrossRef] [PubMed]

67. Culig, Z. Androgen Receptor Coactivators in Regulation of Growth and Differentiation in Prostate Cancer. *J. Cell. Physiol.* **2015**, *231*, 270–274. [CrossRef] [PubMed]

68. Obinata, D.; Takayama, K.; Takahashi, S.; Inoue, S. Crosstalk of the Androgen Receptor with Transcriptional Collaborators: Potential Therapeutic Targets for Castration-Resistant Prostate Cancer. *Cancers* **2017**, *9*, 22. [CrossRef] [PubMed]

69. Moss, P.E.; Lyles, B.E.; Stewart, L.V. The PPARgamma ligand ciglitazone regulates androgen receptor activation differently in androgen-dependent versus androgen-independent human prostate cancer cells. *Exp. Cell Res.* **2010**, *316*, 3478–3488. [CrossRef] [PubMed]

70. Yang, J.P.; Reddy, T.R.; Truong, K.T.; Suhasini, M.; Wong-Staal, F. Functional interaction of Sam68 and heterogeneous nuclear ribonucleoprotein K. *Oncogene* **2002**, *21*, 7187–7194. [CrossRef] [PubMed]

71. Cox, J.; Mann, M. MaxQuant enables high peptide identification rates, individualized p.p.b.-range mass accuracies and proteome-wide protein quantification. *Nat. Biotechnol.* **2008**, *26*, 1367–1372. [CrossRef] [PubMed]

72. Tyanova, S.; Temu, T.; Sinitcyn, P.; Carlson, A.; Hein, M.Y.; Geiger, T.; Mann, M.; Cox, J. The Perseus computational platform for comprehensive analysis of (prote)omics data. *Nat. Methods* **2016**, *13*, 731–740. [CrossRef] [PubMed]

73. Cline, M.S.; Smoot, M.; Cerami, E.; Kuchinsky, A.; Landys, N.; Workman, C.; Christmas, R.; Avila-Campilo, I.; Creech, M.; Gross, B.; et al. Integration of biological networks and gene expression data using Cytoscape. *Nat. Protoc.* **2007**, *2*, 2366–2382. [CrossRef] [PubMed]

74. Lopez-Romero, P. Pre-processing and differential expression analysis of Agilent microRNA arrays using the AgiMicroRna Bioconductor library. *BMC Genom.* **2011**, *12*, 64. [CrossRef] [PubMed]

75. Davis, S.; Meltzer, P.S. GEOquery: A bridge between the Gene Expression Omnibus (GEO) and BioConductor. *Bioinformatics* **2007**, *23*, 1846–1847. [CrossRef] [PubMed]

76. Huang da, W.; Sherman, B.T.; Lempicki, R.A. Systematic and integrative analysis of large gene lists using DAVID bioinformatics resources. *Nat. Protoc.* **2009**, *4*, 44–57. [CrossRef] [PubMed]

International Journal of
Molecular Sciences

Article

BRCA1 Interacting Protein COBRA1 Facilitates Adaptation to Castrate-Resistant Growth Conditions

Huiyoung Yun [1,2,†], Roble Bedolla [1], Aaron Horning [3], Rong Li [3,4], Huai-Chin Chiang [3], Tim-H Huang [3,4], Robert Reddick [5], Aria F. Olumi [6], Rita Ghosh [1,2,3,4] and Addanki P. Kumar [1,2,3,4,7,*]

[1]   Department of Urology, The University of Texas Health, San Antonio, TX 78229, USA; Huiyoung_Yun@DFCI.HARVARD.EDU (H.Y.); BEDOLLAR@uthscsa.edu (R.B.); ghoshr@uthscsa.edu (R.G.)
[2]   Pharmacology, the University of Texas Health, San Antonio, TX 78229, USA
[3]   Molecular Medicine, The University of Texas Health, San Antonio, TX 78229, USA; Horning@livemail.uthscsa.edu (A.H.); lir3@uthscsa.edu (R.L.); chiangh@uthscsa.edu (H.-C.C.); huangt3@uthscsa.edu (T.-H.H.)
[4]   UT Health San Antonio Cancer Center, the University of Texas Health, San Antonio, TX 78229, USA
[5]   Pathology, the University of Texas Health, San Antonio, TX 78229, USA; REDDICK@uthscsa.edu
[6]   Department of Urology, Massachusetts General Hospital Harvard Medical School, 55 Fruit Street, Yawkey Building, Suite 7E, Boston, MA 02215, USA; AOLUMI@PARTNERS.ORG
[7]   South Texas Veterans Health Care System, San Antonio, TX 78229, USA
*   Correspondence: kumara3@uthscsa.edu; Tel.: +1-210-562-4116
†   Present address: Department of Pediatric Oncology, Dana-Farber Cancer Institute, Harvard Medical School, 450 Brookline Ave, Boston, MA 02142, USA.

Received: 31 May 2018; Accepted: 12 July 2018; Published: 20 July 2018

Abstract: COBRA1 (co-factor of BRCA1) is one of the four subunits of the negative elongation factor originally identified as a BRCA1-interacting protein. Here, we provide first-time evidence for the oncogenic role of COBRA1 in prostate pathogenesis. COBRA1 is aberrantly expressed in prostate tumors. It positively influences androgen receptor (AR) target gene expression and promoter activity. Depletion of COBRA1 leads to decreased cell viability, proliferation, and anchorage-independent growth in prostate cancer cell lines. Conversely, overexpression of COBRA1 significantly increases cell viability, proliferation, and anchorage-independent growth over the higher basal levels. Remarkably, AR-positive androgen dependent (LNCaP) cells overexpressing COBRA1 survive under androgen-deprivation conditions. Remarkably, treatment of prostate cancer cells with well-studied antitumorigenic agent, 2-methoxyestradiol (2-ME$_2$), caused significant DNA methylation changes in 3255 genes including COBRA1. Furthermore, treatment of prostate cancer cells with 2-ME$_2$ downregulates COBRA1 and inhibition of prostate tumors in TRAMP (transgenic adenocarcinomas of mouse prostate) animals with 2-ME$_2$ was also associated with decreased COBRA1 levels. These observations implicate a novel role for COBRA1 in progression to CRPC and suggest that COBRA1 downregulation has therapeutic potential.

Keywords: COBRA1; NELFB; androgen receptor; CRPC

---

1. Introduction

Prostate cancer (PCA) continues to be the second leading cause of cancer related deaths in men, with the overwhelming majority of deaths due to castration resistant prostate cancer (CRPC) [1]. Given that early stage PCA is dependent on androgen receptor (AR) signaling, androgen-deprivation therapy (ADT) is the standard therapeutic approach for clinical management of PCA. While ADT is effective in regressing tumor growth, the response is transient (12–18 months) leading to cancer relapse. Relapse of cancer following ADT (known as CRPC) occurs because the recurring tumors grow either in the absence of or low concentrations of androgens [2,3]. CRPC is fatal as no effective durable systemic therapy currently exists. Despite castrate levels of androgens, AR signaling is still active under these conditions and human prostate tumors express AR. Reactivation of AR signaling occurs through numerous mechanisms, such as AR amplification, mutation, splice variants, coregulators, inflammatory cytokines, and receptor tyrosine kinases, contribute to the development of CRPC [4]. These data suggest that development and progression of CRPC is complex and involves compensatory signaling networks. Thus, understanding the molecular factors that contribute to progression to CRPC is critical for successful clinical management of PCA. Towards achieving this goal, we identified an unexpected role for activation of cofactor of BRCA1 (COBRA1), a protein traditionally known to be involved in transcription pausing as yet another mechanism potentially contributing to progression to aggressive prostate cancer.

COBRA1 (aka NELF-B) is one of the four subunits of the negative elongation regulatory (NELF) complex originally identified as a BRCA1-interacting protein. COBRA1 prevents transcriptional elongation by stalling RNA polymerase II (RNAPII) at the proximal promoter region [5,6]. Given its ability to repress transcriptional activity of multiple oncogenes such as estrogen receptor alpha (ERα), earlier studies suggested that COBRA1 may play a tumor-suppressor role [7,8]. Emerging evidence suggests paradoxical oncogenic and tumor suppressive roles for COBRA1. For example, human gastrointestinal adenocarcinomas show increased COBRA1 expression and protein levels compared to normal upper gastrointestinal tract implicating a potential oncogenic role for COBRA1 [9]. Recent studies also indicate a developmental role since mouse embryonic fibroblasts from COBRA1-KO animals show reduced proliferation and elevated apoptosis [10,11]. Interestingly, COBRA1 functions as an AR co-activator by virtue of its ability to interact with AR ligand-binding domain (LBD) [12]. Clinically, patients carrying BRCA mutations are at significantly elevated risk for developing metastatic disease and death from PCA [13]. Furthermore, recent studies showed that nearly 20% of prostate cancer patients who carry the BRCA1 biallelic mutation are at risk for developing castrate resistant prostate cancer. More importantly, these patients carrying biallelic inactivation of BRCA2 are responsive to PARP-1 inhibitors further emphasizing the clinical relevance [14]. Recent mouse genetic studies strongly suggest a mutually antagonistic role of COBRA1 and BRCA1 in both mammary gland development and mammary tumorigenesis [15,16]. However, the role of COBRA1 in prostate cancer is largely unknown. Based on these evidence, we tested whether COBRA1 plays a role in prostate pathogenesis either directly or through its regulatory effects on gene expression. Here, we provide first time evidence for an oncogenic role for COBRA1 in human prostate cancer and its potential as a therapeutic target.

2. Results and Discussion

Basal level and expression of COBRA1 were analyzed in a panel of human prostate cancer cell lines, human prostate tumor array comprising of low (<7) and high (≥7) Gleason score (GS) tumors and a commercial cDNA prostate tissue array. We observed (i) elevated mRNA expression of COBRA1 with increasing tumor aggressiveness (Figure 1a); (ii) significantly increased COBRA1 protein levels in high GS tumors compared with low GS tumors (Figure 1b); and (iii) elevated levels of COBRA1 in an advanced mesenchymal phenotype cell line compared with its isogenic epithelial counterpart (Figure 1c). In silico analysis of Oncomine data showed significantly elevated mRNA expression in prostate tumors compared to normal prostate gland (Figure 1d). These data taken together suggest a potential role for COBRA1 in prostate cancer progression.

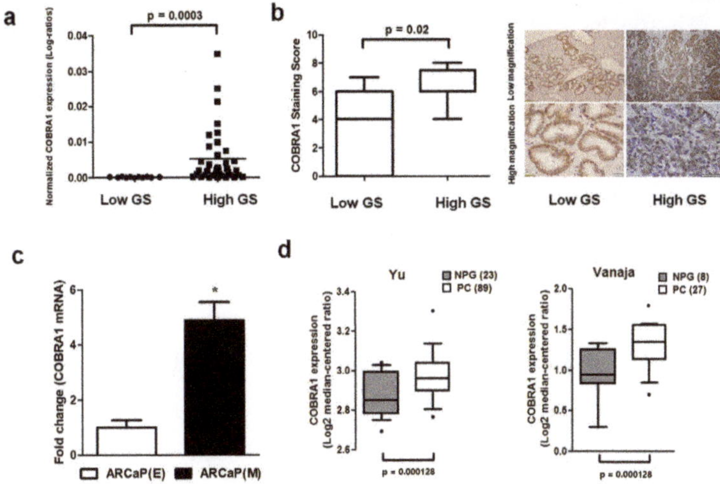

Figure 1. Expression of COBRA1 in human prostate tumors: (**a**) Expression of COBRA1 as assessed by qRT-PCR using the Tissue Scan Prostate Cancer Tissue qPCR Panel III (Origene, Rockville, MD, USA) comprising of low (GS < 7; *n* = 11) and high (GS ≥ 7; *n* = 37) GS tumors. Data was analyzed using unpaired two-tailed *t*-test with Welch's correction (*p* = 0.0003); (**b**) Immunohistochemical evaluation of COBRA1 in human prostate tumor microarray comprising low (GS < 7; *n* = 11) and high (GS ≥ 7; *n* = 13) GS tumors. Cumulative analysis of this data is presented as box plot. Statistical analysis of the data was performed using unpaired two-tailed *t*-test with Welch's correction; (**c**) COBRA1 mRNA expression was measured by qRT-PCR in ARCaP (E) and ARCaP (M) cells. Error bars indicate ±S.E.M. (*n* = 3). * *p* < 0.05; (**d**) Box plots of COBRA1 expression in normal prostate gland (NPG) and prostate carcinoma (PC) from Oncomine database (http://www.oncomine.org, accessed on 19 May 2015). Data sets are log transformed and illustrated as median centered box plots between the differences of mRNA expression within cohorts. Statistical significance was determined by a two-tailed Mann–Whitney test. IHC pictures shown are at 100 and 500 microns (low magnification images) and 100 and 20 microns (high magnification images) for low and high GS tumors respectively.

It was previously shown that COBRA1 interacts with AR LBD and can function as a coactivator of AR [12]. This evidence led to the hypothesis that COBRA1 facilitates androgen independency. To investigate this proposition, AR expressing androgen responsive LNCaP and castrate resistant C4-2B cells with COBRA1 knockdown and overexpression were grown under androgen-deprived conditions. Vector transfected (NTC) and COBRA1 silenced LNCaP cells (shCOBRA1) failed to thrive under these conditions; while COBRA1 overexpressing (pCOBRA1) cells formed large colonies and thrived under androgen-deprived conditions (Figure 2a, left panel). Surprisingly, overexpression or knockdown of COBRA1 had no effect on growth of C4-2B cells under these experimental conditions (Figure 2a, right panel). Stable knockdown of COBRA1 in LNCaP and C4-2B was accompanied by a small but significant reduction in proliferation under hormone-replete conditions (Figure 2b) and COBRA1 overexpression enhanced proliferation in LNCaP but not in C4-2B cells (Figure 2b). These results taken together suggest that COBRA1 may be involved in cellular adaptation under castrated conditions but may not be an important player after cells have adapted to grow in the absence of androgens. To investigate if COBRA1 activates AR signaling, we analyzed AR reporter activity and mRNA expression changes in AR and its bonafide target genes, PSA and TMPRSS2. COBRA1 silenced LNCaP and C4-2B cells had significantly reduced AR-reporter activity in both LNCaP and C4-2B cells (Figure 2c). While silencing COBRA1 significantly reduced AR message levels in LNCaP and C4-2B cells; the AR target genes affected differed between the 2 cells lines. In LNCaP cells TMPRSS2 was significantly reduced while in C4-2B cells PSA (prostate-specific antigen) was significantly reduced (Figure 2d) suggesting differential participation of COBRA1 in AR-mediated transcriptional regulation between androgen-responsive and castrate resistant cells. However, whether subtle differences in the amount of COBRA1 knockdown contributes to the observed differences cannot be ruled out. We interpret these observations to suggest that COBRA1 expression facilitates progression to castrate resistant disease by affecting AR signaling. Our results do not rule out the role for other nuclear receptors in mediating these effects. Further, COBRA1 can physically interact with other transcription factors including Sp1 or Sp3 as there is precedence for interaction of COBRA1 with c-Fos and AP-1 [17]. Our study sets the stage for additional work to understand the mechanism(s) of COBRA1 involvement in prostate cancer progression.

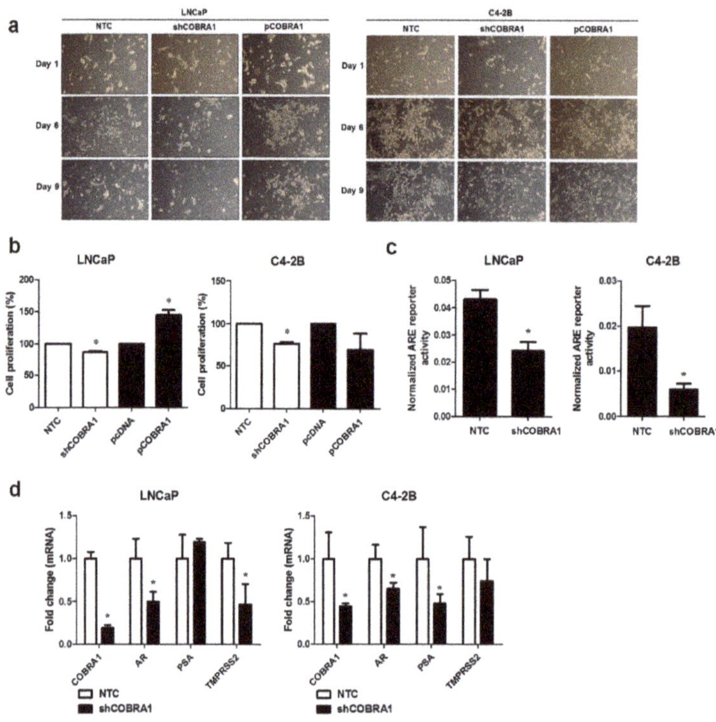

Figure 2. COBRA1 facilitates progression to castrate resistance: (**a**) Androgen responsive LNCaP and its castrate resistant sub line C4-2B stably silenced or ectopically expressing COBRA1 with respective controls were grown in charcoal stripped media for 10 days. Cells were observed microscopically for any morphological changes. A representative image at 10× magnification from three independent experiments is shown; (**b**) Proliferative ability of androgen responsive LNCaP and its castrate resistant sub line C4-2B stably silenced or ectopically expressing COBRA1. Data presented is an average of three independent experiments conducted in triplicate. Error bars indicate ±S.E.M. (n = 3). * p < 0.05; (**c**) ARE reporter activity in androgen responsive LNCaP and its castrate resistant sub line C4-2B stably silenced for COBRA1. Data presented is an average of three independent experiments conducted in triplicate. Error bars represent ±S.E.M. (n = 3). * p < 0.05; (**d**) mRNA expression changes of AR, PSA, and TMPRSS2 in androgen responsive LNCaP and its castrate resistant sub line C4-2B stably silenced for COBRA1. Data presented is an average of three independent experiments conducted in triplicate. Error bars indicate ± S.E.M. (n = 3). * p < 0.05.

There was no significant difference in COBRA1 message among nontransformed and various prostate cancer cell lines although protein level was higher in cancer cells compared with nontransformed cells (Figure 3a,b). Based on protein levels we chose to examine the biological effects of COBRA1 modulation using BPH1 (overexpression), LNCaP, and DU145 (silencing) cells. We observed consistent overexpression (~2 fold) in BPH1-C (BPH1-COBRA1) cells and ~0.5 fold knockdown of COBRA1 in LNCaP and DU145 cells (Figure S1a). Overexpression of COBRA1 resulted in enhanced anchorage independent growth in BPH1 cells, while silencing COBRA1 resulted in decreased anchorage independent growth in DU145 (highest basal COBRA1 level) and had no significant change in LNCaP cells (Figure 3ci–ciii). It is noteworthy to mention that although BPH1 cells exhibited significant increase in anchorage-independent growth, these cells grew slower than the cancer cells, perhaps an indication of their nontumorigenic nature. Similar effects were observed on cell viability with COBRA1 modulation (Figure S1b).

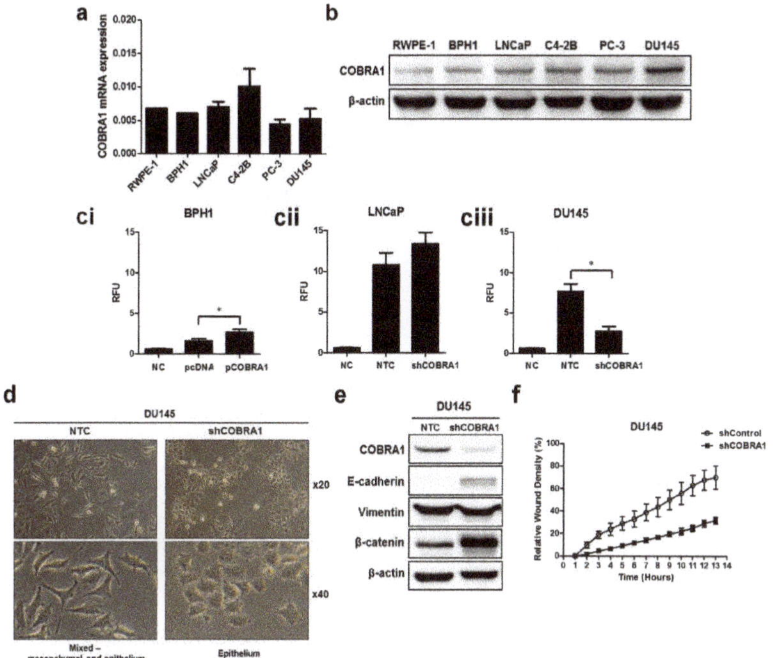

Figure 3. Involvement of COBRA1 in epithelial plasticity: (**a**,**b**) Total RNA and whole cell lysates prepared from logarithmically growing RWPE-1, BPH1, LNCaP, C4-2B, PC-3, and DU145 was used in measuring mRNA expression and protein levels of COBRA1 respectively. In immunoblot analysis β-actin was used as loading control; (**c**) Anchorage-independent growth assay in (**ci**) nontumorigenic BPH1 cells ectopically expressing COBRA1 (pCOBRA1), vector control (pcDNA), or negative control without cells (NC); (**cii**) LNCaP cells and (**ciii**) DU145 cells silenced for COBRA1 (shCOBRA1) or a scrambled shRNA (NTC). Data presented is an average of three independent experiments conducted in triplicate. Statistical analysis of the data was calculated using student's t-test. Error bars indicate ±S.D. (n = 3). * $p < 0.05$; (**d**) Logarithmically growing androgen independent DU145 stably silenced COBRA1 cells with respective controls were observed microscopically for morphological changes. Representative image is shown; (**e**) Immunoblot analysis of E-cadherin, Vimentin and β-catenin in DU145 cells silenced for COBRA1. A representative immunoblot from three independent experiments is shown; (**f**) Migratory ability of DU145 cells silenced for COBRA1. Data shown is a representative of three independent experiments. Error bars indicate ± S.D. (n = 3).

Examination of the morphology of DU145-shCOBRA1 cells showed distinct changes suggestive of epithelial phenotype compared with the non-targeted shRNA transfected cells (NTC) cells that appeared to have a mix of mesenchymal and epithelial phenotype (Figure 3d). This observation prompted us to examine the proteins that are well established markers of epithelial–mesenchymal transition (EMT). We found increased levels of E-cadherin and β-catenin with no changes in vimentin (Figure 3e). These observations are consistent with the data presented in Figure 1c showing that ARCaP-M (mesenchymal cells) have significantly higher expression of COBRA1 than ARCaP-E (epithelial) cells. These data lead us to believe that high levels of COBRA1 in DU145 cells may be associated with cell plasticity due to the lack of E-cadherin/β-catenin complex that play important roles in epithelial barrier. Since gain of cell migration and loss of cell adhesion is a characteristic of mesenchymal cells, we used real-time cell imaging migration assay to test whether COBRA1 silencing would affect the migratory capability. We found significantly decreased migration of shCOBRA1-DU145 cells as a function of time compared with the NTC cells (Figure 3f).

The data presented thus far shows that COBRA1 is overexpressed in prostate tumors and contributes to the adaptation and survival of prostate cancer cells under castrate conditions. To examine whether COBRA1 could serve as a therapeutic target, we analyzed protein changes in the prostate and tumors samples obtained from a retrospective 2-ME$_2$ intervention study conducted in transgenic adenocarcinoma of the mouse prostate (TRAMP) model. We previously demonstrated that 2-ME$_2$ (i) intervention regressed prostate tumor growth in this model and (ii) down regulates c-FLIP [18–20]. Analyses of COBRA1 protein levels showed significant decrease in the prostate from 2-ME$_2$ intervention group compared to the vehicle control (Figure 4a). Consistent with these in vivo observations, treatment with 2-ME$_2$ decreased COBRA1 protein levels in DU145 cells in a dose-dependent manner (Figure 4b). 5 μM 2-ME$_2$ treatment decreased migration of DU145 cells significantly as a function of time (Figure 4c). These results suggest that 2-ME$_2$ could suppress migratory ability of prostate cancer cells in part via inhibition of COBRA1. Furthermore, treating castrate resistant C4-2B cells with 2-ME$_2$ (3 μM) caused significant ($p < 0.05$) DNA methylation changes in 3,255 genes ($n = 91$ hypermethylated and $n = 3164$ hypomethylated) including COBRA1 according to results obtained with an Infinium HumanMethylation450 BeadChip Kit. Functional annotation charts using the Database for Annotation and Visualization and Integrated Discovery (DAVID) on the 3000 most hypomethylated (negative fold change) genes revealed pathways associated with transcription and transcriptional regulation (Figure 4d). For the 1 μM treatment, only the genes exhibiting hypermethylation were run in DAVID together whereas for the 3 μM treatment, only the genes exhibiting hypomethylation were run in DAVID together. Bar charts indicating the level of significance of the association of the DAVID ontology terms with each treatment groups' list of differentially methylated genes (Figure 4d). Of note, hypermethylated genes produced the chart for cells treated with 1 μM 2-ME$_2$ because almost all of the methylation changes observed were positive fold changes. Conversely, treatment with 3 μM 2-ME$_2$ produced hypomethylated genes because most of the methylation changes observed in this treatment group were negative fold changes. Interestingly, we identified COBRA1 as one of the hypomethylated genes in these pathways. Although comprehensive investigations are necessary to conclude whether 2-ME$_2$-mediated decreased expression in COBRA1 is indeed due to changes in its methylation status, nonetheless, these results suggest that 2-ME$_2$ suppresses prostate tumorigenesis possibly by altering the methylation status of COBRA1. This could explain many of the previous observations regarding changes in gene expression in response to 2-ME$_2$ observed by various groups. While this study does not demonstrate the involvement of 2-ME$_2$ in transcriptional pausing, it would be interesting to test the hypothesis that 2-ME$_2$ inhibits COBRA1-mediated RNA Pol II transcriptional activity to prevent prostate pathogenesis. While this manuscript was under preparation, an oncogenic role for the negative elongation factor E (NELFE) was identified in hepatocellular carcinoma [21].

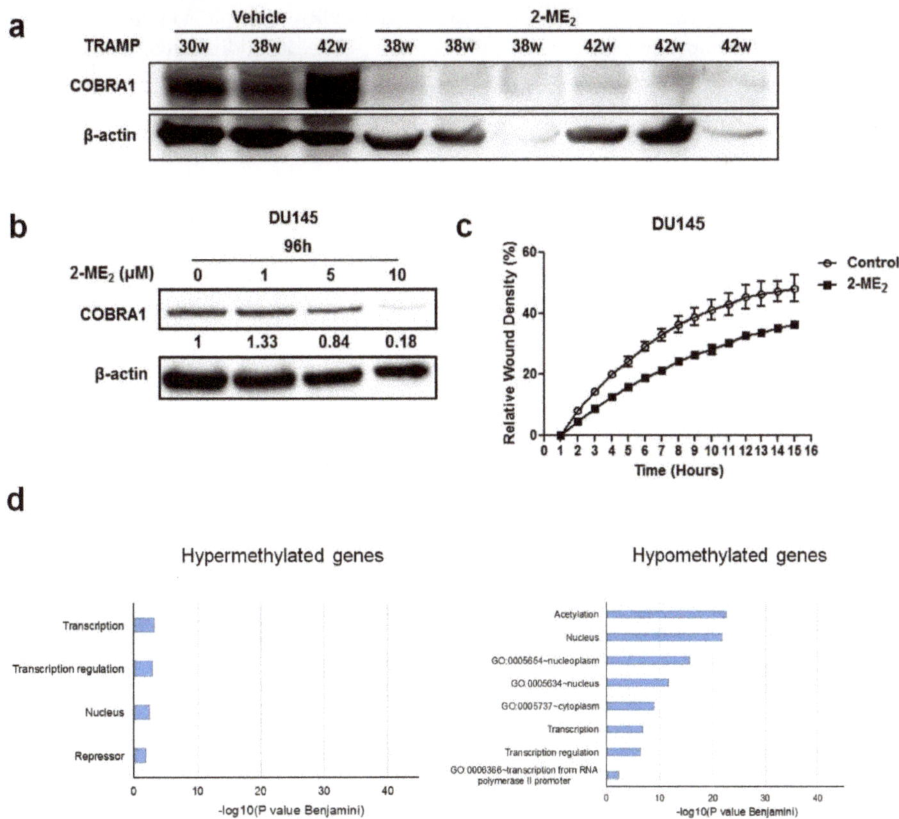

Figure 4. Pharmacological inhibition of COBRA1 reduces migratory ability of DU145 cells and prostate tumor progression: (**a**) Western blot analysis of COBRA1 in the prostate from TRAMP treated with or without 2-ME₂; (**b**) Western blot analysis of COBRA1 in the DU145 cells treated with or without 2-ME₂ (1, 5 and 10 µM) for 24 h; (**c**) Migratory ability of DU145 cells treated with or without 2-ME₂ (5 µM). Data shown is a representative of three independent experiments. Error bars indicate ± S.D. (*n* = 3); (**d**) DNA methylation changes to COBRA1-related transcription-regulation pathways are affected by 2-ME₂ in a concentration dependent manner. Statistically significant gene ontology term associations are indicated by bars ≥1.3.

Although localized prostate cancer can be effectively treated, options for treatment of metastatic castrate resistant disease (CRPC) are mostly palliative with no cure and is therefore a major clinical challenge. Although the treatment landscape for management of CRPC has changed significantly over the past decade, still the pathways that activate AR signaling in the absence or low levels of androgens is poorly defined. These observations underscore the need to understand the cellular, biochemical, and molecular alterations associated with pathological progression to castrate resistance. Along these lines, data presented in this manuscript that show COBRA1 as a potential factor contributing to progression to castrate resistance are significant. To the best of our knowledge these data for the first time implicate oncogenic role of COBRA1 in prostate cancer progression through its ability to allow adaptation to castrate-resistant growth conditions and the loss of epithelial barrier integrity. We also provide evidence that COBRA1 may be a novel therapeutic target in prostate cancer management since treatment with anti-estrogenic compound(s) inhibits COBRA1-related effects observed in prostate cancer. Furthermore, emerging evidence links germline and somatic mutations in DNA repair genes including BRCA1 with castrate resistance [22]. Given that COBRA1 is a BRCA1 interacting protein, we speculate that therapeutic targeting of COBRA1 could provide an additional option for patients with DNA repair aberrations.

3. Materials and Methods

3.1. Cell Culture and Reagents

AR-positive androgen dependent (LNCaP), AR-negative androgen independent (PC-3 and DU145) human prostate cancer cells were purchased from American Type Culture Collection (Manassas, VA, USA), and AR-positive androgen independent (C4-2B) were obtained from Dr. Thambi Dorai (Department of Biochemistry and Molecular Biology, New York Medical College, Valhalla, NY, USA), and benign prostate cells (BPH1) were obtained from Dr. M.S. Lucia (Department of Pathology, University of Colorado Denver, Denver, CO, USA). BPH1, LNCaP, C4-2B, and DU145 cells were maintained in RPMI-1640 media supplemented with 10% fetal bovine serum (FBS) and 1% penicillin/streptomycin in a humidified incubator supplied with 5% CO_2 and at 37 °C. PC-3 cells were grown in F12-K media containing 10% FBS plus antibiotics. Logarithmically growing LNCaP, PC-3, C4-2B (3 μM), and DU145 (5 μM) cells were treated with 2-methoxyestradiol (2-ME$_2$) (Sigma, St. Louis, MO, USA).

3.2. COBRA1 Stable Cell Generation

COBRA1 stable knockdown cells were generated with shRNA targeting COBRA1 using pSUPER-retro-neo retroviral shRNA expression plasmid (Oligoengine, Seattle, WA, USA). In parallel, control cells were generated using a scrambled shRNA. The optimal concentration of G418 (neomycin) for selection and maintenance of COBRA1 stable cells was established by performing kill curve using range of G418 concentrations (0.1–2.0 mg/mL). Cells were seeded at a density of 1×10^5 cells/mL in complete media in T75 flask. Following their attachment, cells were transfected with 10 μg of total plasmid DNA per flask using Lipofectamine 2000 reagent (15 μL; Invitrogen, Grand Island, NY, USA). G418 selection was used to select transfected cells 48–72 h post-transfection (BPH1, 1000 μg/mL; LNCaP, C4-2B, DU145, 500 μg/mL). G418 was replaced every 2–3 days by adding fresh media containing appropriate dose of G418, and cells were examined visually for toxicity daily. Cells were maintained in the media containing G418 and collected as a polyclonal line. The polyclonal cells were plated sparsely at a very low density (~10 cells/well) in 6-well plate and allowed to form individual colonies. The individual colonies were trypsinized and transferred to 10 cm dish for monoclonal expansion. The efficiency of COBRA1 overexpression or knockdown was verified using western blotting and qRT-PCR. However, we noted that the knockdown efficiency decreases with time. For ectopic expression, cells were transfected with pcDNA3.1-based expression vectors for COBRA1 or mock transfected with the empty vector.

3.3. Western Blot and Quantitative Real Time PCR

Whole-cell extracts were prepared using 2× SDS buffer supplemented with fresh protease and phosphatase inhibitors. Equal volumes of protein were fractionated by SDS-PAGE and transferred to nitrocellulose membranes. The primary antibodies used include anti-COBRA1, 1:1000, (Dr. Rong Li, Department of Molecular Medicine, University of Texas Health San Antonio, San Antonio, TX, USA), β-actin, 1:2000, (Sigma, St. Louis, MO, USA), E-cadherin, 1:1000, (Cell signaling, Danvers, MA, USA), Vimentin, 1:1000, (Cell signaling), β-catenin, 1:2000, (Cell signaling). Bound antibody was visualized using ECL kit (Thermo Fisher Scientific, Waltham, MA, USA). All the blots were stripped and reprobed with β-actin to ensure equal loading of protein. Images were captured and analyzed using Gene snap software (Syngene, Frederick, MD, USA), and quantification was carried out using Gene tools software (Syngene).

Total cellular RNA was isolated using Trizol reagent (Invitrogen). RNA was reverse transcribed using the SuperScript VILO cDNA Synthesis Kit (Invitrogen). Target genes were amplified and expression was measured using 7300 Applied Biosystems with SYBR Green dye. The qRT-PCR was conducted with the primers as follows:

β-actin,	forward 5′-GGCACCCAGCACAATGAAGATCAA-3′
	reverse 5′-TAGAAGCATTTGCGGTGGACGATG-3′
COBRA1,	forward 5′-GTTCCAGACAGAGAATGGTG-3′
	reverse 5′-ATACCGACTGGTGGAACT-3′
AR,	forward 5′-AGGAGGAAGGAGAGGCTTCC-3′
	reverse 5′-GAGCAAGGCTGCAAAGGAGT-3′
TMPRSS2,	forward 5′-TACTCTGGAAGTTCATGGGCAGCA-3′
	reverse 5′-AAGTTTGGTCCGTAGAGGCGAACA-3′
PSA,	forward 5′-AATCGATTCCTCAGGCCAGGTGAT-3′
	reverse 5′-AGAACTCCTCTGGTTCAATGCTGC-3′

PCR reactions were conducted in triplicate, and relative mRNA expression was normalized to β-actin. Fold change in experiments was determined relative to solvent control group. Specific amplification of target genes was validated using a dissociation curve.

3.4. Luciferase Assay

For transfections, human prostate cancer cells were plated in triplicate at a density of 1×10^5 cells per well in 24-well plates. Following their attachment, cells were transfected with ARE reporter plasmids (0.5 µg) along with Renilla luciferase (10 ng) using Lipofectamine 2000 reagent (Invitrogen,). Luciferase activity was determined after 36 h transfection using the Dual Luciferase Reporter Assay system (Promega, Madison, WI, USA) essentially as described previously [20].

3.5. Cell Growth and Proliferation

Trypan blue, soft agar growth, and MTT assays were used to determine growth and survival. For trypan blue assay, cells were plated at a density of 1×10^4 cells/well in 24-well plates for 2–3 days, and then trypsinized, combined with the Trypan blue reagent (Sigma, St. Louis, MO, USA), and cell numbers were counted. For soft agar assay, cells were seeded at a density of 10,000 cells/well in 96 well plate containing semisolid agar media. Transformation ability of these cells was measured using CytoSelect™ 96-well Cell Transformation Assay (Cell Biolabs, San Diego, CA, USA) following 6–8 days incubation. Fluorescence was read on SpectraMax M5 plate reader (Molecular Devices, San Jose, CA, USA) using 485/520 nm. Briefly, cells growing in semisolid agar media were solubilized, lysed, and incubated with CyQuant GR Dye (Cell Biolabs) for measuring fluorescence. For cell proliferation, cells were seeded in triplicate at a density of 4×10^3 per well in 96-well plate. Cell proliferation was detected following 72 h incubation essentially as described previously [20] by measuring absorbance at 570/650 nm.

3.6. Migration Assay

The migration rate of androgen independent prostate cancer cells was assessed using the real-time cell imaging system (IncuCyte™ live-cell ESSEN BioScience Inc., Ann Arbor, MI, USA). A scratch was made using the 96-pin WoundMaker™ (ESSEN BioScience Inc.) in cells growing in 96 well plate. Cell migration was monitored in real time over a period of 14 h, and images were automatically acquired and analyzed using IncuCyte™ 96-well Cell Migration Software Application Module (ESSEN BioScience Inc.). Data is represented as the Relative Wound Density (RWD), which is a representation of the spatial cell density in the wound area relative to the spatial cell density outside of the wound area at every time point (time-curve).

3.7. DNA Methylation Array

Global DNA methylation levels in androgen independent prostate cancer cells C4-2B were measured following treatment with 2-ME$_2$ (1 µM and 3 µM) for 5 days. By using a MethylMiner Methylated DNA enrichment kit (Invitrogen), methylated DNA was isolated from fragmented whole genomic DNA via binding to the methyl-CpG binding domain of human MBD protein coupled to paramagnetic Dynabeads M-280 Streptavidin through a biotin linker. Then, samples were subjected to DNA methylation analysis on the Illumina HumanMethylation450 Beadchip Kit (BASIC core facility at UTHSA). This analysis produced a list of genes with significant changes in DNA methylation (hyper or hypomethylation; $p < 0.05$). For each gene list, up to 3000 genes were selected and run in the Database for Annotation and Visualization and Integrated Discovery (DAVID) that produced Functional Annotation Charts.

3.8. Animal Experiments

Western blot analysis of COBRA1 levels were examined in TRAMP (transgenic adenocarcinomas of mouse prostate), prostate tumors and tissues were obtained from a repository available from previous studies in our laboratory [18,19]. Tissues were procured from a study testing the potential of 2-ME$_2$ (50 mg/kg body weight through drinking water) by administrating to 22–25 week old of TRAMP mice for additional 25 weeks [18,19]. Proteins were extracted using RIPA buffer from the tumors and tissues in the control group (30, 38, and 42 weeks), and in the treatment group (38 and 42 weeks).

3.9. Immunohistochemistry

COBRA1 rabbit polyclonal antibody (Dr. Rong Li, Department of Molecular Medicine, University of Texas Health San Antonio, San Antonio, TX, USA) was used. Sections from paraffin embedded tissues were heat cleared and rehydrated. Antigen retrieval was performed with citrate buffer at pH 6.0 in a 121 °C pressure chamber. Endogenous peroxidase was quenched with a TBS buffer containing 3% hydrogen peroxide followed by a protein blocking buffer incubation. Each step was carried out at room temperature. The sections were incubated for 1 h at room temperature with the antibody. The negative control sections were incubated with a Universal Rabbit negative control Rabbit Ig fraction (DAKO Corp., Carpinteria, CA, USA). The ancillary and visualization systems were: Rabbit HRP polymer (BioCare Medical, Concord, CA, USA) and DAB Chromogen System (DAKO Corp.). IHC slides were evaluated and graded by pathologist (R. R) in a blinded fashion. The total COBRA1 staining was scored as the product of the staining intensity (on a scale of 0–3) and the percentage of cells stained (on a scale of 0–5), resulting in a scale of 0–8. Staining intensity was scored as follows: 0, none of the cells scored positively; 1, weak staining; 2, moderate staining intensity; and 3, strong staining intensity. Percent staining was scored as follows: 1, 20%; 2, 30%; 3, 60%; 4, 80% and 5, 100% cells stained. Low (100 µm) and high (50 µm) magnification images were taken using a Nikon Eclipse Ci microscope equipped with camera (D5-F12).

3.10. Oncomine Data

COBRA1 expressed in normal prostate gland and prostate carcinoma were obtained from two independent studies for each gene expression in the Oncomine database. Primary sources are from different group's microarray data mentioned in the graph (http://www.oncomine.org). Data sets are log transformed and illustrated as median centered box plots between the differences of mRNA transcription within cohorts. Statistical significance was determined by a two-tailed Mann–Whitney test. Detailed information of the standardized normalization and statistical calculations are indicated on the Oncomine website.

3.11. Statistical Analysis

All numerical results are expressed as mean ± S.D. or S.E.M. derived from 3 independent experiments, unless otherwise stated. Statistical analyses were conducted using Student's *t*-test and statistically significant differences were established as $p < 0.05$. The statistical significance of IHC data was calculated using unpaired two-tailed t test with a Welch's correction.

Supplementary Materials: Supplementary materials can be found at http://www.mdpi.com/1422-0067/19/7/2104/s1.

Author Contributions: Conceptualization, A.P.K. and R.G.; Methodology, H.Y., A.H., and R.B.; Validation, A.P.K. and H.Y.; Formal Analysis, H.Y., R.B., and A.H.; Investigation, H.Y., R.B.; Resources, A.P.K., T.-H.H., R.L., and H.-C.C.; Data Curation, H.Y.; Writing-Original Draft Preparation, H.Y.; Writing-Review & Editing, A.P.K., R.G., H.Y., R.L., T.-H.H., R.B., R.L., R.R., H.-C.C., A.H., and A.F.O.; Visualization, H.Y., R.B., R.R., and A.H.; Supervision, A.P.K.; Project Administration, A.P.K.; Funding Acquisition, A.P.K. Authorship must be limited to those who have contributed substantially to the work reported.

Funding: This research was supported in part by funds from Cancer Prevention Research Institute of Texas (CPRIT RP 150166) National Cancer Institute R01 CA135451; National Center for Complementary and Alternative Medicine AT 005513-01A1, AT007448-01; Veterans Affairs-Merit Award 1 I01 BX 000766-01 & BX 003876 (APK); R01 CA149516 (RG); R01 CA220578 (RL) and a postdoctoral fellowship to H-C. C from NIH (T32CA148724). APK acknowledge support provided by the CTRC 40th Anniversary Distinguished Professor of Oncology Endowment. We acknowledge support provided by Bioanalytics and Single-Cell Core (BASiC) core facility at UT Health San Antonio and UT Health San Antonio Cancer Center through the National Cancer Institute support grant #2P30 CA 054174-17 (APK and RG).

Conflicts of Interest: The authors declare no conflicts of interest.

Abbreviations

COBRA1	Co-factor of BRCA1
2-ME$_2$	2-methoxyestradiol
TRAMP	Transgenic adenocarcinoma of the mouse prostate
PCA	Prostate cancer
CRPC	Castration resistant prostate cancer
AR	Androgen receptor
ADT	Androgen-deprivation therapy
NELF	Negative elongation factor
RNAPII	RNA polymerase II
ERα	Estrogen receptor alpha
LBD	Ligand binding domain
EMT	Epithelial-mesenchymal transition
DAVID	Database for annotation and visualization and integrated discovery

References

1. Siegel, R.L.; Miller, K.D.; Jemal, A. Cancer statistics, 2015. *CA Cancer J. Clin.* **2015**, *65*, 5–29. [CrossRef] [PubMed]
2. Chang, S.S. Treatment options for hormone-refractory prostate cancer. *Rev. Urol.* **2007**, *9*, S13. [PubMed]
3. Cetin, K.; Li, S.; Blaes, A.H.; Stryker, S.; Liede, A.; Arneson, T.J. Prevalence of patients with non metastatic prostate cancer on androgen deprivation therapy in the United States. *Urology* **2013**, *81*, 1184–1189. [CrossRef] [PubMed]
4. Yamaoka, M.; Hara, T.; Kusaka, M. Overcoming persistent dependency on androgen signaling after progression to castration-resistant prostate cancer. *Clin. Cancer Res.* **2010**, *16*, 4319–4324. [CrossRef] [PubMed]
5. Yamaguchi, Y.; Takagi, T.; Wada, T.; Yano, K.; Furuya, A.; Sugimoto, S.; Hasegawa, J.; Handa, H. NELF, a multisubunit complex containing RD, cooperates with DSIF to repress RNA polymerase II elongation. *Cell* **1999**, *97*, 41–51. [CrossRef]
6. Ye, Q.; Hu, Y.F.; Zhong, H.; Nye, A.C.; Belmont, A.S.; Li, R. BRCA1-induced large-scale chromatin unfolding and allele-specific effects of cancer-predisposing mutations. *J. Cell Biol.* **2001**, *155*, 911–921. [CrossRef] [PubMed]
7. Sun, J.M.; Spencer, V.A.; Li, L.; Chen, H.Y.; Yu, J.; Davie, J.R. Estrogen regulation of trefoil factor 1 expression by estrogen receptor alpha and Sp proteins. *Exp. Cell Res.* **2005**, *302*, 96–107. [CrossRef] [PubMed]

8. Aiyar, S.E.; Blair, A.L.; Hopkinson, D.A.; Bekiranov, S.; Li, R. Regulation of clustered gene expression by cofactor of BRCA1 (COBRA1) in breast cancer cells. *Oncogene* **2007**, *26*, 2543–2553. [CrossRef] [PubMed]

9. McChesney, P.A.; Aiyar, S.E.; Lee, O.J.; Zaika, A.; Moskaluk, C.; Li, R. Cofactor of BRCA1: A novel transcription factor regulator in upper gastrointestinal adenocarcinomas. *Cancer Res.* **2006**, *66*, 1346–1353. [CrossRef] [PubMed]

10. Amleh, A.; Nair, S.J.; Sun, J.; Sutherland, A.; Hasty, P.; Li, R. Mouse cofactor of BRCA1 (Cobra1) is required for early embryogenesis. *PLoS ONE* **2009**, *4*, e5034. [CrossRef] [PubMed]

11. Williams, L.H.; Fromm, G.; Gokey, N.G.; Henriques, T.; Muse, G.W.; Burkholder, A.; Fargo, D.C.; Hu, G.; Adelman, K. Pausing of RNA polymerase II regulates mammalian developmental potential through control of signaling networks. *Mol. Cell* **2015**, *58*, 311–322. [CrossRef] [PubMed]

12. Sun, J.; Blair, A.L.; Aiyar, S.E.; Li, R. Cofactor of BRCA1 modulates androgen-dependent transcription and alternative splicing. *J. Steroid Biochem. Mol. Biol.* **2007**, *107*, 131–139. [CrossRef] [PubMed]

13. Robinson, D.; Van Allen, E.M.; Wu, Y.M.; Schultz, N.; Lonigro, R.J.; Mosquera, J.M.; Montgomery, B.; Taplin, M.E.; Pritchard, C.C.; Attard, G.; et al. Integrative clinical genomics of advanced prostate cancer. *Cell* **2015**, *161*, 1215–1228. [CrossRef] [PubMed]

14. Mateo, J.; Carreira, S.; Sandhu, S.; Miranda, S.; Mossop, H.; Perez-Lopez, R.; Nava Rodrigues, D.; Robinson, D.; Omlin, A.; Tunariu, N.; et al. DNA-repair defects and olaparib in metastatic prostate cancer. *New Engl. J. Med.* **2015**, *373*, 1697–1708. [CrossRef] [PubMed]

15. Nair, S.J.; Zhang, X.; Chiang, H.C.; Jahid, M.J.; Wang, Y.; Garza, P.; April, C.; Salathia, N.; Banerjee, T.; Alenazi, F.S.; et al. Genetic suppression reveals DNA repair-independent antagonism between BRCA1 and COBRA1 in mammary gland development. *Nat. Commun.* **2016**, *7*, 10913. [CrossRef] [PubMed]

16. Zhang, X.; Chiang, H.-C.; Wang, Y.; Zhang, C.; Smith, S.; Zhao, X.; Nair, S.J.; Michalek, J.; Jatoi, I.; Lautner, M.; et al. Attenuation of RNA Polymerase II Pausing Mitigates BRCA1-Associated R-loop Accumulation and Tumorigenesis. *Nat. Commun.* **2017**, *8*, 15908. [CrossRef] [PubMed]

17. Zhong, H.; Zhu, J.; Zhang, H.; Ding, L.; Sun, Y.; Huang, C.; Ye, Q. COBRA1 inhibits AP-1 transcriptional activity in transfected cells. *Biochem. Biophys. Res. Commun.* **2004**, *325*, 568–573. [CrossRef] [PubMed]

18. Garcia, G.E.; Wisniewski, H.G.; Lucia, M.S.; Arevalo, N.; Slaga, T.J.; Kraft, S.L.; Strange, R.; Kumar, A.P. 2-Methoxyestradiol inhibits prostate tumor development in transgenic adenocarcinoma of mouse prostate: Role of tumor necrosis factor-alpha-stimulated gene 6. *Clin. Cancer Res.* **2006**, *12*, 980–988. [CrossRef] [PubMed]

19. Ganapathy, M.; Ghosh, R.; Jianping, X.; Xiaoping, Z.; Bedolla, R.; Schoolfield, J.; Yeh, I.T.; Troyer, D.A.; Olumi, A.F.; Kumar, A.P. Involvement of FLIP in 2-methoxyestradiol-induced tumor regression in transgenic adenocarcinoma of mouse prostate model. *Clin. Cancer Res.* **2009**, *15*, 1601–1611. [CrossRef] [PubMed]

20. Yun, H.; Xie, J.; Olumi, A.F.; Ghosh, R.; Kumar, A.P. Activation of AKR1C1/ER induces apoptosis by downregulation of c-FLIP in prostate cancer cells: A prospective therapeutic opportunity. *Oncotarget* **2015**, *6*, 11600–11613. [CrossRef] [PubMed]

21. Dang, H.; Takai, A.; Forgues, M.; Pomyen, Y.; Mou, H.; Xue, W.; Ray, D.; Ha, K.C.; Morris, Q.D.; Hughes, T.R.; et al. Oncogenic activation of the RNA binding protein NELFE and MYC signaling in hepatocellular carcinoma. *Cancer Cell* **2017**, *32*, 101–114. [CrossRef] [PubMed]

22. Mateo, J.; Boysen, G.; Barbieri, C.E.; Bryant, H.E.; Castro, E.; Nelson, P.S.; Olmos, D.; Pritchard, C.C.; Rubin, M.A.; de Bono, J.S. DNA Repair in Prostate Cancer: Biology and Clinical Implications. *Eur. Urol.* **2017**, *71*, 417–425. [CrossRef] [PubMed]

International Journal of
Molecular Sciences

Article

High Throughput Chemical Screening Reveals Multiple Regulatory Proteins on FOXA1 in Breast Cancer Cell Lines

Shixiong Wang [1,*], Sachin Kumar Singh [1], Madhumohan R. Katika [1], Sandra Lopez-Aviles [2] and Antoni Hurtado [1,*]

[1] Breast Cancer Research group, Nordic EMBL Partnership, Centre for Molecular Medicine Norway (NCMM), University of Oslo, P.O. 1137 Blindern, 0318 Oslo, Norway; parmarsachinsingh@gmail.com (S.K.S.); maddycdfd@gmail.com (M.R.K.)

[2] Cell Cycle Research group, Nordic EMBL Partnership, Centre for Molecular Medicine Norway (NCMM), University of Oslo, P.O. 1137 Blindern, 0318 Oslo, Norway; s.l.aviles@ncmm.uio.no

* Correspondence: shixiong.wang@ncmm.uio.no (S.W.); toni.hurtado@ncmm.uio.no (A.H.); Tel.: +47-22-845-865 (S.W.); +47-22-840-774 (A.H.)

Received: 30 October 2018; Accepted: 14 December 2018; Published: 19 December 2018

Abstract: Forkhead box A1 (FOXA1) belongs to the forkhead class transcription factor family, playing pioneering function for hormone receptors in breast and prostate cancers, and mediating activation of linage specific enhancers. Interplay between FOXA1 and breast cancer specific signaling pathways has been reported previously, indicating a regulation network on FOXA1 in breast cancer cells. Here in this study, we aimed to identify which are the proteins that could potentially control FOXA1 function in breast cancer cell lines expressing different molecular markers. We first established a luciferase reporter system reflecting FOXA1 binding to DNA. Then, we applied high throughput chemical screening of multiple protein targets and mass spectrometry in breast cancer cell lines expressing different molecular markers: ER positive/HER2 negative (MCF-7), ER positive/HER2 positive (BT474), and ER negative/HER2 positive (MDA-MB-453). Regardless of estrogen receptor status, HER2 (human epidermal growth factor receptor 2) enriched cell lines showed similar response to kinase inhibitors, indicating the control of FOXA1 by cell signaling kinases. Among these kinases, we identified additional receptor tyrosine kinases and cyclin-dependent kinases as regulators of FOXA1. Furthermore, we performed proteomics experiments from FOXA1 immunoprecipitated protein complex to identify that FOXA1 interacts with several proteins. Among all the targets, we identified cyclin-dependent kinase 1 (CDK1) as a positive factor to interact with FOXA1 in BT474 cell line. In silico analyses confirmed that cyclin-dependent kinases might be the kinases responsible for FOXA1 phosphorylation at the Forkhead domain and the transactivation domain. These results reveal that FOXA1 is potentially regulated by multiple kinases. The cell cycle control kinase CDK1 might control directly FOXA1 by phosphorylation and other kinases indirectly by means of regulating other proteins.

Keywords: breast cancer; FOXA1; drug screening and proteomics

---

1. Introduction

Forkhead Box A1 (FOXA1), also known as hepatocyte nuclear factor 3 alpha, belongs to the forkhead family of transcription factors and plays pivotal roles in the development of prostate and mammary gland [1,2]. FOXA1 is a pioneer transcription factor with a DNA binding domain that resembles the structure of linker histones [3,4]. Moreover, the C-terminal domain interacts with the core histones to endorse chromatin opening [5]. Genome-wide studies have shown that FOXA1 is enriched at regions with enhancer histone marks H3K4me1 and H3K4me2 [6] and that mediates the activation of lineage enhancers [7]. Recently, it has been reported that FOXA1 recruits histone methyl transferase to mediate the methylation on H3K4me1 [8]. FOXA1 is also able to control the expression of TET1 and interact with it to regulate local DNA methylation and the activity of corresponding enhancer activity [9]. These properties confer the ability of FOXA1 to act as a pioneer factor for both estrogen receptor α [10,11] and androgen receptor [12], controlling their binding, location, and function in breast and prostate cancer respectively.

Breast cancer is the most common cancer type among women in developed countries. Breast cancer is a heterogeneous disease with several subtypes showing differences in histopathology, tumor biology, and prognosis. Despite of such complexity, around 70% of breast cancers cases are estrogen receptor alpha (ER) positive. ER is a nuclear receptor that mediates the response to estrogen and triggers a transcription program driving the proliferation of cancer cells [13,14]. Therefore, ER and its associated metabolism have been the targets of endocrine therapies, which have significantly improved the survival rate of ER positive breast cancer patients [15]. However, resistance to endocrine therapies has been observed in a substantial fraction of ER positive patients. The resistance is often associated with gain of receptor tyrosine kinase function [15] such as ERBB2/HER2. In addition, ER somatic mutations can also result in hormone resistance by activating the receptor in the absence of ligand binding in metastatic ER positive tumors [16,17]. Interestingly, almost all the ER binding chromatin interactions are dependent on the pioneer factor FOXA1, a dependence that is even preserved in hormone resistant tumors [10]. ER and FOXA1 are co-expressed in metastatic endocrine resistant tumor and the redistribution of ER binding correlates with FOXA1 binding [18]. Moreover, overexpression of FOXA1 also mediates endocrine resistance by varying the ER-regulated transcripts and the IL-8 signaling in preclinical model [19]. These particular properties support the idea that FOXA1 could be an attractive target of ER positive breast cancer especially in endocrine resistant context. However, targeting directly a pioneer factor might provoke undesired side effects in healthy tissues. On the other hand, targeting proteins with a role activating FOXA1 could be a plausible alternative. Hence, in this study, we aimed to investigate which proteins can control FOXA1 in different breast cancer cell lines. For this aim, we took a high throughput chemical screening approach (with known protein targets) in order to search for proteins controlling FOXA1 in breast cancer cells. As readout, we used a luciferase reporter system, which is able to reflect FOXA1 binding to DNA. After two rounds of high throughput chemical screening, we identified several interesting proteins that could be potential FOXA1 regulators. Finally, by means of performing proteomics experiments we could identify that cyclin-dependent kinases 1 (CDK1) might directly regulate FOXA1 by phosphorylation.

2. Results

2.1. Generation of the FOXA1 Luciferase Reporter System

In order to perform the high throughput screening, we constructed a luciferase reporter system reflecting the binding of FOXA1 to forkhead motifs. The promoter of the TFF1 gene, which contains two forkhead motifs, was cloned upstream of the luciferase expression cassette of pGL4.20 plasmid (Figure 1A). Previously, it was demonstrated that with a similar construct the luciferase expression was controlled by the FOXA1 binding to the promoter of the chosen gene [20]. To validate the system, we transfected the reporter plasmid into MCF-7 (ER positive/HER2 negative), BT474 (ER positive/HER2 positive), and MDA-MB-453 (ER negative/HER2 positive) breast cancer cell lines. We choose MCF-7 and BT474 cell lines is because they are positive for the expression of ER. The cell line MDA-MB-453 here serves as an ER negative control cell line. The luciferase assay was performed 48 h after transfection. In MCF-7, the reporter showed a much higher activity than the empty vector control, which means that the promoter can drive the expression of the downstream luciferase reporter (Figure 1B). To further validate that the luciferase signal was FOXA1 dependent, the two-forkhead motifs in the promoter were mutated. Hence, the corresponding created mutants were named BS1 (binding site 1 mutated), BS2 (binding site 2 mutated) and BS1/2 (both 1 and 2 mutated). In MCF-7, the mutation of BS1 almost abolished the reporter activity to the level of the empty vector control, while the BS2 only reduced the signal around 20% (Figure 1B). The data suggested that the BS1 site played the main role in mediating FOXA1 binding. The double mutant BS1/2 also showed a significant reduction of luciferase signal in all cell lines tested (Figure 1B,C), with a level similar to BS1 in MCF-7. Moreover, knockdown of FOXA1 with siRNA in MCF-7 also abolished the activity of the WT reporter (Figure 1D), which confirmed the FOXA1 specificity of the reporter system. Previously, it has been reported that both ER and FOXA1 bind to TTF1 promoter and induce its expression [20]. Importantly, our experiments were performed in estrogen-depleted conditions, which is a condition that impedes the binding of ER to the TFF1 promoter. In Figure S1 the binding of ER and FOXA1 at the promoter of TFF1-LUC construct is shown. We performed chromatin immunoprecipitation (ChIP) of both transcription factors in MCF7 cells transfected with the TFF1-LUC-WT and TFF1-LUC-mutant (BS1/2) in estrogen-depleted cells followed by real-time PCR. The experiment reveals that FOXA1 interacts to the promoter of TFF1-LUC construct but ER interaction is almost undetectable. Moreover, the data shown in Figure S1 demonstrate that FOXA1 binds to the promoter of TFF1-LUC-WT vector and that the binding disappears in cells carrying mutations of the FKH motif (BS1/2). Hence, our data reveals clearly that the luciferase signal is mainly mostly due to FOXA1 binding.

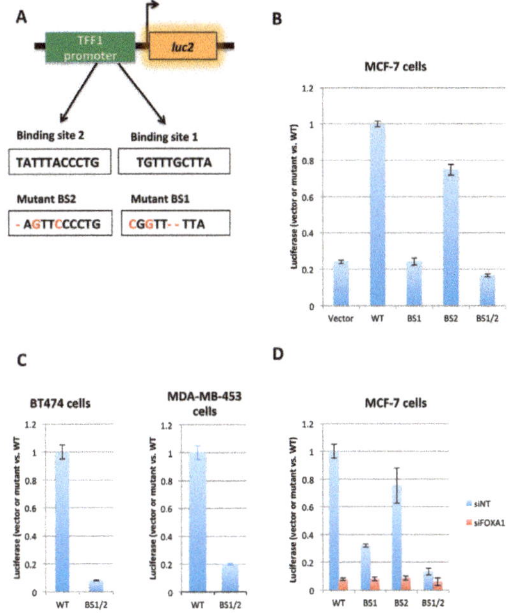

Figure 1. Design of the luciferase reporter and validation. (**A**) Schematic design of the luciferase reporter reflecting FOXA1 activity and the mutations of FOXA1 binding sites. The promoter of Tff1 gene was cloned into the upstream of the luciferase expression cassette of pGL4.20 vector, two FOXA1 motifs are mutated to the sequences in red. (**B**) The wild type reporter (WT), reporter with the mutation in BS1 (BS1), BS2 (BS2), and double mutant (BS1/2) were transfected into MCF-7 cells for validation. Luciferase assay was performed to measure the reporter activity (*n* = 3). (**C**) Wild type and double mutant reporter plasmids were validated further with BT474 (**left**) and MDA-MB-453 (**right**) cell lines (*n* = 3). (**D**) The pGL4.20-WT, BS1, BS2, and BS1/2 were transfected into MCF-7 together with non-targeting siRNA (siNT) and siRNA targeting FOXA1 (siFOXA1). Luciferase assay was performed 48 h after transfection (*n* = 3).

2.2. Multiple Targets Were Identified as Potential FOXA1 Regulators

To test the hypothesis that FOXA1 could be regulated by multiple kinases/proteins, we performed a high throughput chemical screening with the reporter system constructed above. The screening pipeline is illustrated in Figure S2. Briefly, the luciferase reporter was transfected into all MCF-7, BT474, and MDA-MB-453 breast cancer cell lines overnight. Then, cells were re-plated into 384 well plates and maintained in DMEM media free of hormones overnight. Cells were treated with chemicals from a drug library (Enzo Life Sciences; http://www.enzolifesciences.com/) at 10μM concentration. A total of 550 drugs (Table S1) were used in the screening and the luciferase assay was performed 24 h after the start of chemical treatment.

The data from the chemical screening was analyzed, and drugs with a significant impact were selected based on the fold change of the luciferase signal (T test comparing control treated vs. treated with drug; p-value < 0.05 was used as cut-off). We considered uniquely drugs that resulted in an increase of at least 50% or in a reduction of at least 40% of the luciferase signal. Based on the idea that luciferase signal correlates with FOXA1 binding to the TTF1 promoter, the compounds that provide a gain of luciferase signal might be targeting proteins that inhibit FOXA1 binding to the chromatin. By contrast, those compounds that provide a loss of luciferase signal might be targeting proteins that stimulate FOXA1 binding to the chromatin. The numbers of total drugs with a significant impact on luciferase signal from each cell line are summarized in Figure 2A. In MCF-7 cells (HER2 negative), more chemicals that increased luciferase signal were identified compared to inhibitory chemicals (55 and 35, respectively). BT474 cells (HER2 positive) showed a different behavior with fewer chemicals increasing the luciferase signal compared to inhibitory chemicals (14 and 136, respectively). Moreover, the analysis of the other HER2 positive cell line (MDA-MB-453) revealed similar results as the ones obtained with BT474 cells (19 and 144 chemicals that increased or decreased luciferase signal, respectively). Next, we evaluated which were the proteins targeted by the chemicals identified in our screening and which group of proteins they were enriched for each of the cell lines investigated. Our results revealed that the highest fraction of the compounds were kinase inhibitors (Figure 2B). Interestingly, the fraction of compounds was greater in HER2 positive cells (52% in BT474 and 64% in MDA-MB-453) compared to HER2 negative (41% in MCF-7). This finding correlates with the activity of signaling pathways (kinases) in the corresponding cell lines. Due to the high HER2 level in BT474 and MDA-MB-453, cellular signaling pathways in these two cell lines are highly active. By contrast, MCF-7 cell growth is mainly dependent on ER induced transcription, and signaling pathways are not as active as in HER2 positive cell lines. Hence, our data suggests that additional active kinases in signaling pathways may contribute to FOXA1 activity in cell lines with high HER2. Interestingly when we analyzed the overlap between the different cell lines we observed that BT474 and MDA-MB-453 shared most of their inhibitory chemicals (Figure 2C). Altogether, these results indicate a high similarity of kinases that potentially regulate FOXA1 in these two HER2 positive cell lines regardless of ER status.

2.3. Second Screening Narrowed down the Number of Compound Target Candidates

In order to increase the specificity of the screening and narrow down the number of positive drugs (and their respective targets) for functional validation, a second round of chemical screening was performed using fewer chemicals and lower concentrations. We were more interested in targets that activate FOXA1 and thus only inhibitory drugs from the first screening were selected. In addition, considering that most of the inhibitory chemicals were kinase inhibitors, we performed an in silico phosphorylation prediction using Group-based Prediction System 3.0 (GPS 3.0) [21], in order to identify potential phosphorylation sites in FOXA1. The result of the analysis showed that multiple sites in FOXA1 are potential phosphorylation sites for different kinases. By comparing the in silico phosphorylation analysis and the targets of positive chemicals from the screening (Figure 3A), a list of 45 chemicals were selected for the second round of screening at 5 and 1 μM concentrations using MCF-7, BT474, and MDA-MB-453 cell lines.

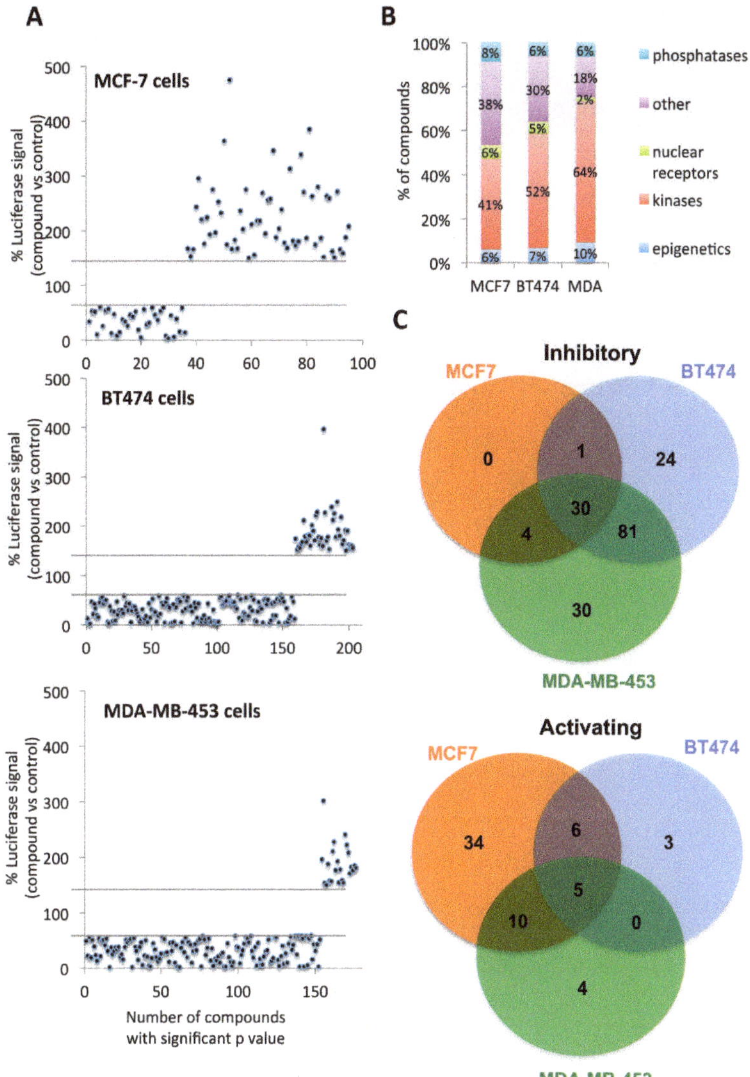

Figure 2. Multiple chemicals were identified positive in the chemical screening. (**A**) Three plots indicating the number of significant compounds (*T* test; two tails; $p < 0.05$) that affect the luciferase expression in each of the breast cancer cell lines investigated (MCF-7, BT474, and MDA-MB-453). Each plot illustrates the % of luciferase expression of cells treated with compounds and normalized to control treated cells (treatment/control). We have represented the compounds with a significant increase (more than 150%) or decrease (less than 40%) luciferase expression compared to control. (**B**) Fraction (expressed in %) of significant compounds targeting different group of proteins: phosphatases, nuclear receptors, kinases, epigenetics and other groups. The plot represents the % of group of compounds with a significant p value for each cell line investigated. (**C**) Venn-diagram showing the overlap of positive chemicals between MCF-7, BT474, and MDA-MB-453 cells. Inhibitory (**upper**) and activating (**lower**) are showed independently. The number of positive chemicals in MCF-7, BT474, and MDA-MB-453 were showed in different columns with activating chemicals in red and inhibitory chemicals in blue.

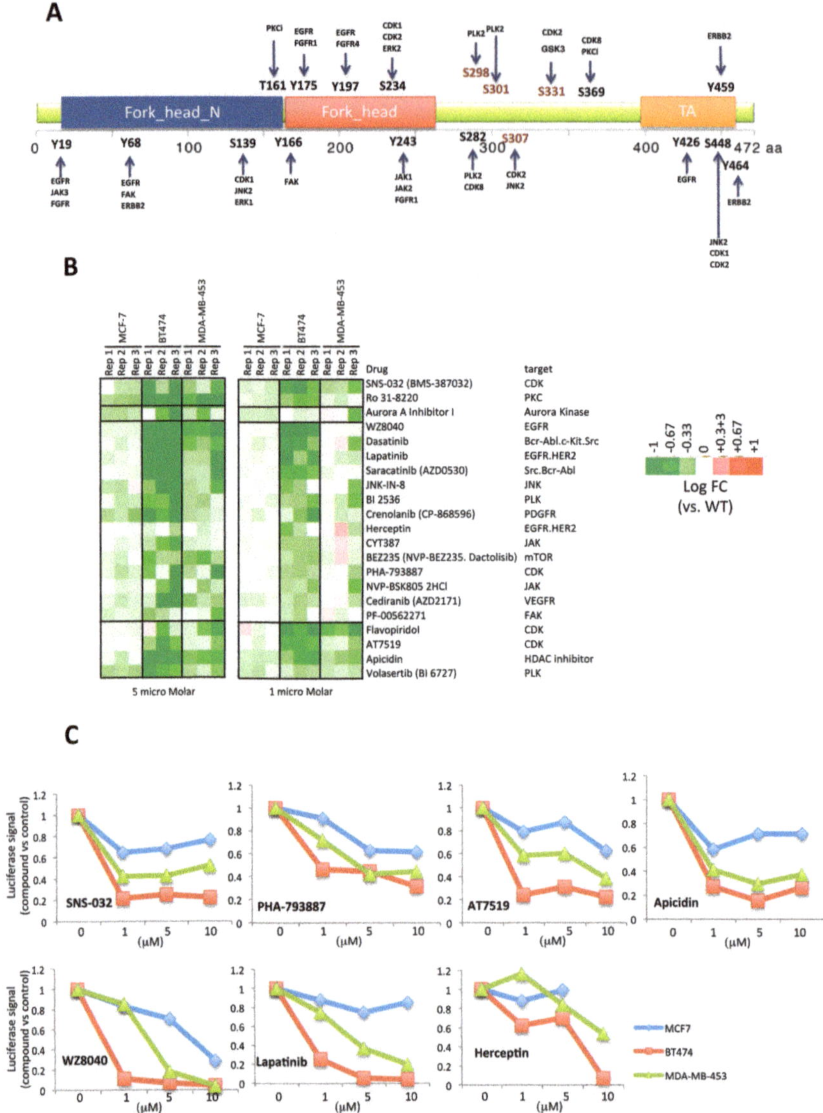

Figure 3. Validation of chemicals by the second screening. (**A**) Diagram showing potential FOXA1 phosphorylation sites and corresponding kinases. The prediction of FOXA1 phosphorylation sites was performed with GPS 3.0, and corresponding kinases that overlapped with positive targets in the first chemical screening were identified. Both overlapping kinases and their potential sites are labeled. (**B**) Heatmap showing the result of compounds with a significant change in luciferase expression. 45 chemicals were used for the second chemical screening for three cell lines and two concentrations (5 and 1 microM). The heatmap illustrates the log2 fold change in luciferase expression (drug treatment vs. control) of cells treated with the 21 compounds with a significant p value. (**C**) Dose responses of compounds with a significant inhibitory effect in the expression of luciferase. The plots represent the relative signal of the compounds targeting the receptor tyrosine kinases HER2/EGFR and the cyclin-dependent kinases (CDK) vs. control treated cells. The average of three independent experiments for different concentrations tested (0, 1, 5, and 10 micro Molar) is plotted.

The results with the lower concentration used (1 µM) revealed that 21 chemicals were still positive as inhibitors of the reporter. Moreover, some of these compounds were cell type specific (Figure 3B,C, Figures S3 and S4). The results of the screening have identified interesting targets, some of them that have already been shown to be associated with hormone resistance. In this regard, we have identified two inhibitors targeting CDK and PKC in all cell lines. A few chemicals were still shared by BT474 and MDA-MB-453 (4 chemicals), while more than half of all the chemicals (14 of 21 chemicals) were mostly targeting proteins in BT474 cells. These chemicals targeted CDK, receptor tyrosine kinases (EGFR, VEGFR, PDGFR), PLK, JAK, and other intracellular kinases such as PKC, JNK. Finally, an Aurora kinase inhibitor was mainly identified as a positive drug in MCF-7 cells.

2.4. FOXA1 Pulldown and Proteomics Identify CDK1 as a Potential Direct Regulator

Taken together, data above showed that several kinases might control positively FOXA1 function. Moreover, our data suggests that more kinases impinge on FOXA1 function/activity in HER2 positive cell lines. Next, we aimed to identify whether any of the kinases identified from our drug screening might be directly regulating FOXA1. Hence, we used proteomic approaches to identify protein targets of FOXA1 for breast cancer cells. For that, we performed mass spectrometry in MCF-7 and BT474 cells from chromatin immunoprecipitation extracts by using a specific FOXA1 antibody. This resulted in 116 and 139 proteins identified to be interacting with FOXA1 in MCF-7 and BT474 cells, respectively (Table S2). Importantly, this proteomic approach allowed us to identify that around 28% (32 proteins) of the FOXA1 pulled down proteins in MCF-7 were also identified within the FOXA1 pull down in BT474, which confirms the suitability of this method to identify FOXA1-interacting proteins (Figure 4A). In addition, we have also identified other FOXA1 proteins likely to be specific for each cell type investigated. Among these targets identified in BT474 in the chemical screening, we have found CDK1 as a FOXA1 protein partner in the proteomics experiment (Figure 4B).

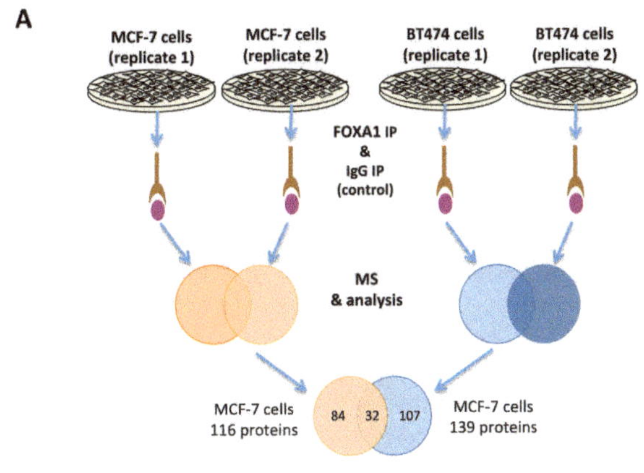

B

Protein	Number of peptides in MCF-7			Number of peptides in BT474		
	Rep 1	Rep 2	IgG	Rep 1	Rep 2	IgG
CDK1 (Cyclin Dependent Kinase 1)	0	0	0	5	1	0
HDAC7 (Histone deacetylase 7)	1	1	0	0	0	0

Figure 4. Proteomics results from FOXA1 immunoprecipitation (IP). (**A**) Workflow for FOXA1 IP proteomics. Two independent replicates were performed and uniquely proteins without peptides identified at IgG control were considered positive. Moreover, we considered exclusively proteins identified in both of the replicates. The figure includes a Venn diagram that compares the number of FOXA1 interacting proteins shared between MCF-7 and BT474 cells and the ones identified exclusively each cell line tested. (**B**) Peptide enrichment of CDK1 and HDAC7 in both cell lines is included. The rest of peptide enrichment of the other identified proteins can be found at supplementary information.

3. Discussion

This work has revealed that several drugs targeting kinases influence the binding of FOXA1 to our TTF1-LUC reporter cassette. Interestingly, the findings from the drug screening indicate that additional kinases have a positive control in HER2 positive cells compared to MCF-7 cells. These results suggest the hypothesis that the binding of FOXA1 to the chromatin in HER2 enriched cells is induced by additional kinases compared to HER2 negative cell lines. The increased number of kinases that positively regulate FOXA1 in HER2 positive cells might be associated with a gain of binding to chromatin. Another possible interpretation of that finding might be understood as a mechanism by which HER2 positive cells guarantee the FOXA1 binding to the chromatin by means of increasing the number of kinases that positively regulate FOXA1. In this regard, it has been recently reported that PI3K might influence the binding of FOXA1 to the chromatin by regulating the activity of the methyltransferase enzyme MLL2 [22]. These enzymes can directly methylate histone H3 on position 4, which are in fact the epigenetic mark recognized by FOXA1 in order to interact with chromatin. Hence, the inhibition of MLL2 by PI3K impacts negatively in the binding of FOXA1 to the chromatin. These results might be in contradiction to our findings, which suggest that the binding of FOXA1 to the DNA might be positively regulated by the same kinase. One possibility of such discrepancies might be due to PI3K inhibiting the subset of FOXA1 chromatin binding regions enriched with DNA sequence motifs for ER interaction. Such hypothesis is supported by the fact that the treatment of PI3K increases the sensitivity of breast cancer cells to anti-estrogen drugs. Hence, the inhibition of PI3K might facilitate indirectly the binding of ER by allowing the binding of FOXA1 and therefore the pioneer function of FOXA1 to ER enriched regions as it has been already reported [23]. Moreover, these results are not in disagreement with the fact that the binding of FOXA1 to additional chromatin regions might be also induced by PI3K/mTOR kinase. In this regard, our drug screening data suggest that mTOR might impact FOXA1 indirectly by other mechanisms that do not imply FOXA1 phosphorylation. In this regard, in our drug screening we have identified that mTOR inhibitors have a negative impact in the binding of FOXA1 to chromatin. Interestingly, it has been reported that the pharmacologic inhibition of GSK3 antagonizes the suppressive effects on the growth of mTOR inhibitors [24,25], which suggests that the kinase GSK3 might be operating downstream of mTOR. Given these evidences, together with the results of this study showing that Ser 331 of FOXA1 contains a consensus site for GSK3 phosphorylation, we might hypothesize that mTOR inhibition affects FOXA1 binding to DNA through the regulation of GSK3. Whether this phosphorylation confers an increase of activity of FOXA1 needs to be elucidated yet.

In this study, we have observed that CDK1 is a protein that interacts with FOXA1 in HER2 positive cells. The interplay between FOXA1 and CDK1 has not being investigated previously. In this regard, it has been reported that CDK1 can regulate gene transcription at S phase of the cell cycle [26]. In particular, CDK1 phosphorylates key transcription factors in S phase and regulates their activity and protein stability [26]. Moreover, a previous study [27] has reported that FOXA1 binds to chromatin in mitosis. This study has also reported that the FOXA1 mitotic binding helps cells to prepare for the transcriptional reactivation of interphase genes after mitotic exit. Importantly, CDK1 activity increases during mitosis, leading to the phosphorylation of proteins whose function is required during this phase of the cell cycle. At the metaphase to anaphase transition, degradation of cyclin B results in a drop in CDK1 activity and mitotic exit. Considering that FOXA1 interacts with CDK1 in BT474 cells but not in MCF-7 cells, one might hypothesize that FOXA1 might be regulated by CDK1 to prepare cells for the transcriptional reactivation of genes just after mitosis. Our in silico analyses for putative sites of phosphorylation have identified several CDK consensus motifs in FOXA1. Moreover, we have observed that the treatment with several general CDK inhibitors have a negative impact on FOXA1 binding to chromatin. Interestingly, this effect was stronger in BT474 compared to the other cell lines investigated. Altogether it is reasonable to postulate that FOXA1 is a potential substrate of CDK1, and the mitotic chromatin binding of FOXA1 could be regulated by CDK1 phosphorylation. If the hypothesis is correct, inhibitors repressing CDK activity should influence the affinity between FOXA1 and its target binding motifs, as observed for the TFF1 reporter. Future experiments might confirm whether FOXA1 is a substrate of CDK1 and should also resolve how such phosphorylation impacts its function.

4. Materials and Methods

4.1. Cell Culture

MCF-7, BT-474, and MDA-MB-453 cell lines were purchased from American Type Culture Collection (ATCC, Manassas, VA, USA). MCF-7 and MDA-MB-453 cell lines were cultured in in DMEM (4.5 g/L glucose) supplemented with 10% FBS BT474 was cultured in DMEM (4.5 g/L glucose) supplemented with 10% FBS, and 0.01 mg/mL insulin.

4.2. Construction of the Luciferase Reporter System

pGL4.20 plasmid was purchased from Promega (E6751). The sequence of wild type TFF1 promoter was obtained from UCSC genome browser (http://genome.ucsc.edu/) with the genomic coordinate chr21:43,786,510-43,787,509 of the GRCh37/hg19 assembly. The promoter of TFF1 gene was cloned into the pGL4.20 with Mul I and Bgl II restriction sites. FOXA1 binding sites mutagenesis was performed with QuikChange II Site-Directed Mutagenesis Kit (200524, Agilent, Santa Clara, CA, USA) following the manufacture's protocol.

4.3. Transfection and Luciferase Assay

MCF-7, BT474, and MDA-MB-453 cells were plated into 96 well plate in full culture media. 24 h after transfection, pGL4.20-TFF1-Pro-WT (with wild type FOXA1 binding sites) and pGL4.20-TFF1-Pro-BS1, BS2, and BS1/2 (containing corresponding mutation in FOXA1 binding sites) were transfected into all cell lines with Lipofectamine 2000 (Invitrogen, Carlsbad, CA, USA) following the manufacture's protocol. pRL-TK Renilla luciferase control reporter plasmid was co-transfected as the control for transfection efficiency. siRNA targeting FOXA1 (ON-TARGET J-010319-05-0005, Thermo Fisher Scientific, Waltham, MA, USA) or siControl Non-targeting (siNT) (SI03650318, Promega, Madison, WI, USA) were co-transfected with the pGL4.20-TFF1 reporter system with Lipofectamine 2000 (Invitrogen). Luciferase assay was directly carried out in the plate 48 h after transfection with the Dual-Glo Luciferase Reporter Assay System (E2920, Promega) following the manufacture's protocol.

4.4. Chromatin Immunoprecipitation

Chromatin immunoprecipitation was performed as described previously [28]. In brief, first MCF-7 cells transfected with pGL4.20-TFF1 luciferase reporter were crosslinked with 1% formaldehyde for 10 min followed by quenching with 125 mM glycine. Then chromatin was sheared with sonication and incubated together with antibodies against FOXA1 (ab23738, Abcam, Cambridge, UK) or estrogen receptor alpha (sc-543, Santa Cruz Biotechnology, Dallas, TX, USA) respectively at 4 °C overnight. Immunoprecipitated DNA was purified with phenol-chloroform extraction followed by ethanol precipitation. FOXA1 and ER binding to the TFF1 luciferase reporter was detected by qPCR with a primer pair ranging across the restriction site used for cloning. The sequences of the qPCR primers are: CACCATGGAGAACAAGGTGA (forward) and AACAGTACCGGATTGCCAAG (reverse).

4.5. High Throughput Chemical Screening and Analysis

MCF-7, BT474, and MDA-MB-453 cells were plated into 10cm culture dish in complete media at 70% confluence. 24 h after plating, 15μg of pGL4.20-TFF1-Pro-WT was transfected with Lipofectamine 2000 following the manufacture's protocol. 5 h after transfection, cells were washed with PBS and media was changed to clear DMEM medium supplemented with 5% stripped serum. After 2 h, transfected cells were plated into 384-well plate (781098, Greiner, Kremsmünster, Austria) in clear DMEM medium supplemented with 5% stripped serum. 24 h after transfection, cells were treated with selected chemical library at a final concentration of 10 μM (5 and 1 μM for the second screening). Three technical replicates were performed for each chemical. 24 h after the chemical treatment, luciferase assay was performed with Steady-Glo luciferase assay system (E2510, Promega) directly in 384-well plate following the manufacture's protocol.

Positive chemicals were selected with Student's *t*-test (*p*-value < 0.05, and with increase >50% or reduction >40% vs. vehicle). Heat map of the second screening was generated with Java TreeView Cluster 3.0 (https://sourceforge.net/projects/jtreeview/).

4.6. FOXA1 Phosphorylation Sites Prediction

Group-based Prediction System 3.0 (GPS 3.0) (http://gps.biocuckoo.org/) was used to predict potential phosphorylation sites in FOXA1 protein. Medium threshold was chosen for the analysis.

4.7. Immunoprecipitation

FOXA1 immunoprecipitation was performed with Pierce Crosslink IP kit (26147, Thermo Fisher Scientific) following the manufacture's protocol. A total of 6 dishes (10 cm) of MCF-7 or BT474 cells cultured in full media were lyzed with Lysis/wash buffer. Cell lysate was loaded to columns containing protein A/G beads that cross-linked to anti-FOXA1 antibody (Abcam, ab23738). The incubation was kept at 4 °C for O/N followed by elusion with the Elusion buffer.

4.8. Protein Digestion

Total 50 μL of an IP protein extract with 200 μL of UA (Urea buffer) in the filter unit and centrifuge at 14,000× *g* for 15 min. Add 200 μL of UA to the filter unit and centrifuge at 14,000× *g* for 15 min. The flow-through was discarded. Next 100 μL IAA solution was added and mixed at 600 rpm in a thermo-mixer for 1 min and incubated without mixing for 20 min. Furthermore, the filter units were centrifuge at 14,000× *g* for 10 min. The filter unit was washed with adding 100 μL of UA and centrifuging at 14,000× *g* for 15 min. This step was repeated twice. Next the filter unit was equilibrated with 100 μL of ABC (ammonium bicarbonate) and centrifuge at 14,000× *g* for 10 min. This step as repeated twice. Finally, protein was digested by adding 40 μL ABC with trypsin (enzyme to protein ratio 1:50) and mix at 600 rpm in thermo-mixer for 1 min and Incubated the units at 37 °C for 18 h. Afterwards, the filter units was transfer to new collection tubes and centrifuged at 14,000× *g* for 10 min. One more time, 40 μL ABC was added and centrifuged at 14,000× *g* for 10 min to recover all digested peptides. Last filtrate was acidified with CF_3COOH.

4.9. Desalting Digested Peptides

We used Oasis HLB cartridges (10 mg) from Waters (product no. 186000383, Oslo, Norway). Briefly, the following steps were followed: (1) condition the HLB cartridges with 1 mL 100% ACN (Acetonitrile); (2) equilibrate with 1.5 mL 2% ACN and 0.1% TFA buffer (wash solution); (3) load sample; (4) wash with 1 mL of wash solution; and (5) elute with 1 mL glycolic acid buffer (1 M glycolic acid, 5% TFA, 80% acetonitrile).

4.10. Data Processing and Analysis

Peptides were analyzed with Q-Exactive mass-spectrometry. Raw mass spectrometric data were analyzed in the MaxQuant tool and employed Andromeda for database search. The mass spectra were matched against the human Uniprot FASTA database. Enzyme specificity was set to trypsin, and the search included cysteine carbamidomethylation as a fixed modification and *N*-acetylation of protein, oxidation of methionine, and/ or phosphorylation of Ser, Thr, Tyr residue (STY) as variable modifications. Up to two missed cleavages were allowed for protease digestion, and peptides had to be fully tryptic.

Supplementary Materials: Supplementary materials can be found at http://www.mdpi.com/1422-0067/19/12/4123/s1.

Author Contributions: All the experiments were conceived by M.R.K., S.W., S.K.S., S.L.-A., and A.H. All the in vitro experiments were conducted by S.W. and S.K.S., Computational analyses were conducted by S.W. and S.K.S. Manuscript was written by A.H. and S.W. with help from the other authors.

Funding: This research was funded by NFR-young talenged grant number [250554/F20].

Acknowledgments: The sequencing service was provided by the Norwegian Sequencing Center (www.sequencing.uio.no), a national technology platform hosted by the University of Oslo and supported by the "Functional Genomics" and "Infrastructure" programs of the Research Council of Norway and the Southeastern Regional Health Authorities. M.R.K. and S.K.S. were supported by NCMM and S.W. and A.H. were supported by NFR. Finally, we would like to thank Elvia Valentini for assisting technically in this work.

Conflicts of Interest: The authors disclose no conflicts of interest.

References

1. Mirosevich, J.; Gao, N.; Matusik, R.J. Expression of Foxa transcription factors in the developing and adult murine prostate. *Prostate* **2005**, *62*, 339–352. [CrossRef] [PubMed]
2. Bernardo, G.M.; Lozada, K.L.; Miedler, J.D.; Harburg, G.; Hewitt, S.C.; Mosley, J.D.; Godwin, A.K.; Korach, K.S.; Visvader, J.E.; Kaestner, K.H.; et al. FOXA1 is an essential determinant of ERα expression and mammary ductal morphogenesis. *Development* **2010**, *137*, 2045–2054. [CrossRef] [PubMed]
3. Cirillo, L.A.; McPherson, C.E.; Bossard, P.; Stevens, K.; Cherian, S.; Shim, E.Y.; Clark, K.L.; Burley, S.K.; Zaret, K.S. Binding of the winged-helix transcription factor HNF3 to a linker histone site on the nucleosome. *EMBO J.* **1998**, *17*, 244–254. [CrossRef] [PubMed]
4. Clark, K.L.; Halay, E.D.; Lai, E.; Burley, S.K. Co-crystal structure of the HNF-3/fork head DNA-recognition motif resembles histone H5. *Nature* **1993**, *364*, 412–420. [CrossRef]
5. Cirillo, L.A.; Lin, F.R.; Cuesta, I.; Friedman, D.; Jarnik, M.; Zaret, K.S. Opening of Compacted Chromatin by Early Developmental Transcription Factors HNF3 (FoxA) and GATA-4. *Mol. Cell* **2002**, *9*, 279–289. [CrossRef]
6. Lupien, M.; Eeckhoute, J.; Meyer, C.A.; Wang, Q.; Zhang, Y.; Li, W.; Carroll, J.S.; Liu, X.S.; Brown, M. FoxA1 Translates Epigenetic Signatures into Enhancer-Driven Lineage-Specific Transcription. *Cell* **2008**, *132*, 958–970. [CrossRef]
7. Sérandour, A.A.; Avner, S.; Percevault, F.; Demay, F.; Bizot, M.; Lucchetti-Miganeh, C.; Barloy-Hubler, F.; Brown, M.; Lupien, M.; Métivier, R.; et al. Epigenetic switch involved in activation of pioneer factor FOXA1-dependent enhancers. *Genome Res.* **2011**, *21*, 555–565. [CrossRef]
8. Jozwik, K.M.; Chernukhin, I.; Serandour, A.A.; Nagarajan, S.; Carroll, J.S. FOXA1 Directs H3K4 Monomethylation at Enhancers via Recruitment of the Methyltransferase MLL3. *Cell Rep.* **2016**, *17*, 2715–2723. [CrossRef]
9. Yang, Y.A.; Zhao, J.C.; Fong, K.W.; Kim, J.; Li, S.; Song, C.; Song, B.; Zheng, B.; He, C.; Yu, J. FOXA1 potentiates lineage-specific enhancer activation through modulating TET1 expression and function. *Nucleic Acids Res.* **2016**, *44*, 8153–8164. [CrossRef]
10. Hurtado, A.; Holmes, K.A.; Ross Innes, C.S.; Schmidt, D.; Carroll, J.S. FOXA1 is a key determinant of estrogen receptor function and endocrine response. *Nat. Genet.* **2011**, *43*, 27–33. [CrossRef]
11. Carroll, J.S.; Liu, X.S.; Brodsky, A.S.; Li, W.; Meyer, C.A.; Szary, A.J.; Eeckhoute, J.; Shao, W.; Hestermann, E.V.; Geistlinger, T.R.; et al. Chromosome-Wide Mapping of Estrogen Receptor Binding Reveals Long-Range Regulation Requiring the Forkhead Protein FoxA1. *Cell* **2005**, *122*, 33–43. [CrossRef] [PubMed]
12. Yang, Y.A.; Yu, J. Current perspectives on FOXA1 regulation of androgen receptor signaling and prostate cancer. *Genes Dis.* **2015**, *2*, 144–151. [CrossRef] [PubMed]
13. Perou, C.M.; Sørlie, T.; Eisen, M.B.; Van De Rijn, M.; Jeffrey, S.S.; Rees, C.A.; Pollack, J.R.; Ross, D.T.; Johnsen, H.; Akslen, L.A.; et al. Molecular portraits of human breast tumours. *Nature* **2000**, *406*, 747–752. [CrossRef] [PubMed]
14. Sørlie, T.; Perou, C.M.; Tibshirani, R.; Aas, T.; Geisler, S.; Johnsen, H.; Hastie, T.; Eisen, M.B.; Van De Rijn, M.; Jeffrey, S.S.; et al. Gene expression patterns of breast carcinomas distinguish tumor subclasses with clinical implications. *Proc. Natl. Acad. Sci. USA* **2001**, *98*, 10869–10874. [CrossRef]
15. Osborne, C.K.; Schiff, R. Mechanisms of Endocrine Resistance in Breast Cancer. *Annu. Rev. Med.* **2011**, *62*, 233–247. [CrossRef]

16. Robinson, D.R.; Wu, Y.-M.; Vats, P.; Su, F.; Lonigro, R.J.; Cao, X.; Kalyana-Sundaram, S.; Wang, R.; Ning, Y.; Hodges, L.; et al. Activating ESR1 mutations in hormone-resistant metastatic breast cancer. *Nat. Genet.* **2013**, *45*, 1446–1451. [CrossRef] [PubMed]

17. Fanning, S.W.; Mayne, C.G.; Dharmarajan, V.; Carlson, K.E.; Martin, T.A.; Novick, S.J.; Toy, W.; Green, B.; Panchamukhi, S.; Katzenellenbogen, B.S.; et al. Estrogen receptor alpha somatic mutations Y537S and D538G confer breast cancer endocrine resistance by stabilizing the activating function-2 binding conformation. *eLife* **2016**, *5*, e12792. [CrossRef]

18. Ross-Innes, C.S.; Stark, R.; Teschendorff, A.E.; Holmes, K.A.; Ali, H.R.; Dunning, M.J.; Brown, G.D.; Gojis, O.; Ellis, I.O.; Green, A.R.; et al. Differential oestrogen receptor binding is associated with clinical outcome in breast cancer. *Nature* **2012**, *481*, 389–393. [CrossRef]

19. Fu, X.; Jeselsohn, R.; Pereira, R.; Hollingsworth, E.F.; Creighton, C.J.; Li, F.; Shea, M.; Nardone, A.; De Angelis, C.; Heiser, L.M.; et al. FOXA1 overexpression mediates endocrine resistance by altering the ER transcriptome and IL-8 expression in ER-positive breast cancer. *Proc. Natl. Acad. Sci. USA* **2016**, *113*, E6600–E6609. [CrossRef]

20. Laganière, J.; Deblois, G.; Lefebvre, C.; Bataille, A.R.; Robert, F.; Giguère, V. From the Cover: Location analysis of estrogen receptor alpha target promoters reveals that FOXA1 defines a domain of the estrogen response. *Proc. Natl. Acad. Sci. USA* **2005**, *102*, 11651–11656. [CrossRef]

21. Wang, B.; Wang, M.; Li, A. Prediction of post-translational modification sites using multiple kernel support vector machine. *PeerJ* **2017**, *5*, e3261. [CrossRef] [PubMed]

22. Toska, E.; Osmanbeyoglu, H.U.; Castel, P.; Chan, C.; Hendrickson, R.C.; Elkabets, M.; Dickler, M.N.; Scaltriti, M.; Leslie, C.S.; Armstrong, S.A.; et al. PI3K pathway regulates ER-dependent transcription in breast cancer through the epigenetic regulator KMT2D. *Science* **2017**, *355*, 1324–1330. [CrossRef] [PubMed]

23. Bosch, A.; Li, Z.; Bergamaschi, A.; Ellis, H.; Toska, E.; Prat, A.; Tao, J.J.; Spratt, D.E.; Viola-Villegas, N.T.; Castel, P.; et al. PI3K inhibition results in enhanced estrogen receptor function and dependence in hormone receptor–positive breast cancer. *Sci. Transl. Med.* **2015**, *7*, 283ra51. [CrossRef] [PubMed]

24. Koo, J.; Yue, P.; Gal, A.A.; Khuri, F.R.; Sun, S.-Y. Maintaining Glycogen Synthase Kinase-3 Activity Is Critical for mTOR Kinase Inhibitors to Inhibit Cancer Cell Growth. *Cancer Res.* **2014**, *74*, 2555–2568. [CrossRef]

25. Thorne, C.A.; Wichaidit, C.; Coster, A.D.; Posner, B.A.; Wu, L.F.; Altschuler, S.J. GSK-3 modulates cellular responses to a broad spectrum of kinase inhibitors. *Nat. Chem. Biol.* **2015**, *11*, 58–63. [CrossRef]

26. Landry, B.D.; Mapa, C.E.; Arsenault, H.E.; Poti, K.E.; Benanti, J.A. Regulation of a transcription factor network by Cdk1 coordinates late cell cycle gene expression. *EMBO J.* **2014**, *33*, 1044–1060. [CrossRef] [PubMed]

27. Caravaca, J.M.; Donahue, G.; Becker, J.S.; He, X.; Vinson, C.; Zaret, K.S. Bookmarking by specific and nonspecific binding of FoxA1 pioneer factor to mitotic chromosomes. *Genes Dev.* **2013**, *27*, 251–260. [CrossRef]

28. Gilfillan, S.; Fiorito, E.; Hurtado, A. Functional genomic methods to study estrogen receptor activity. *J. Mammary Gland Biol. Neoplasia* **2012**, *17*, 147–153. [CrossRef]

International Journal of
Molecular Sciences

Article

Molecular Insights into the Classification of Luminal Breast Cancers: The Genomic Heterogeneity of Progesterone-Negative Tumors

Gianluca Lopez [1,2,†], Jole Costanza [3,†], Matteo Colleoni [1], Laura Fontana [4], Stefano Ferrero [1,5], Monica Miozzo [3,4] and Nicola Fusco [1,5,*]

1   Division of Pathology, Fondazione IRCCS Ca' Granda-Ospedale Maggiore Policlinico, 20122 Milan, Italy; gianluca.lopez@unimi.it (G.L.); matteo.colleoni1@studenti.unimi.it (M.C.); stefano.ferrero@unimi.it (S.F.)
2   School of Pathology, University of Milan, 20122 Milan, Italy
3   Research Laboratory Unit, Fondazione IRCCS Ca' Granda Ospedale Maggiore Policlinico, 20122 Milan, Italy; jole.costanza@policlinico.mi.it (J.C.); monica.miozzo@unimi.it (M.M.)
4   Medical Genetics, Department of Pathophysiology and Transplantation, University of Milan, 20122 Milan, Italy; laura.fontana@unimi.it
5   Pathology, Department of Biomedical, Surgical, and Dental Sciences, University of Milan, 20122 Milan, Italy
*   Correspondence: nicola.fusco@unimi.it; Tel.: +39-02-5503-8425; Fax: +39-02-5503-4243
†   These authors contributed equally to this work.

Received: 20 December 2018; Accepted: 23 January 2019; Published: 25 January 2019

Abstract: Estrogen receptor (ER)-positive progesterone receptor (PR)-negative breast cancers are infrequent but clinically challenging. Despite the volume of genomic data available on these tumors, their biology remains poorly understood. Here, we aimed to identify clinically relevant subclasses of ER+/PR− breast cancers based on their mutational landscape. The Cancer Genomics Data Server was interrogated for mutational and clinical data of all ER+ breast cancers with information on PR status from The Cancer Genome Atlas (TCGA), Memorial Sloan Kettering (MSK), and Molecular Taxonomy of Breast Cancer International Consortium (METABRIC) projects. Clustering analysis was performed using gplots, ggplot2, and ComplexHeatmap packages. Comparisons between groups were performed using the Student's *t*-test and the test of Equal or Given Proportions. Survival curves were built according to the Kaplan–Meier method; differences in survival were assessed with the log-rank test. A total of 3570 ER+ breast cancers (PR− n = 959, 27%; PR+ n = 2611, 73%) were analyzed. Mutations in well-known cancer genes such as *TP53*, *GATA3*, *CDH1*, *HER2*, *CDH1*, and *BRAF* were private to or enriched for in PR− tumors. Mutual exclusivity analysis revealed the presence of four molecular clusters with significantly different prognosis on the basis of *PIK3CA* and *TP53* status. ER+/PR− breast cancers are genetically heterogeneous and encompass a variety of distinct entities in terms of prognostic and predictive information.

Keywords: breast cancer; progesterone receptor negative; mutational profiling; PI3K pathway; TP53

---

1. Introduction

Estrogen receptor (ER)-positive progesterone receptor (PR)-negative (ER+/PR−) breast cancers are a subset of Luminal B tumors characterized by the strong and diffuse nuclear expression of ER-alpha but not of PR [1]. They account for 5% of all invasive breast cancers and show a relatively aggressive clinical course compared to ER+/PR+ neoplasms [1–5]. ER+/PR− invasive breast cancers are described as larger in size than PR+ carcinomas and are generally of no special histological type (i.e., ductal) [1,6]. Even though they preferentially affect postmenopausal women, these diagnoses are not exceptional in younger patients [1,2,7]. As confirmed by several prospectively randomized controlled neoadjuvant trials, ER+/PR− breast cancers are associated with a higher response but also worse long-term outcome after neoadjuvant therapy [5]. There are several lines of evidence to suggest that the worse prognosis of ER+/PR− tumors may be related to the phenomena of hormone therapy resistance [1–5]. However, a large adjuvant trial on the use of aromatase inhibitors in postmenopausal women with early breast cancer revealed that the PR status has no effect on the relative efficacy of this therapy [8]. For this reason, some authors have questioned the clinical utility of PR testing [9]. To date, hormonal therapy remains recommended in ER+ tumors regardless of PR status [10]. All these diverse correlations highlight the clinical challenges provided by ER+/PR− breast cancers.

A proportion of ER+/PR− neoplasms shows a remarkable degree of genomic instability, reaching almost twice the DNA copy number variations and tumor mutational load than those of both ER+/PR+ and ER− breast cancers [1,8]. Furthermore, many growth factors were observed to be overexpressed in these tumors, such as HER family, PI3K, Akt, and src [1,2,11–13]. These pathways, which can also be altered in ER+/PR+ tumors, are known to be involved in ER phosphorylation, which may lead to ligand-independent activation [14]. There is also evidence that the upregulation of Akt and HER1/2 is implicated in tamoxifen resistance [1,2,11,12,15–18]. Recently, PR has been proposed as a surrogate biomarker of altered growth factor signaling [5]. Due to these insights, and the substantial lack of distinct biological properties identified to date in ER+/PR− breast cancers, it is becoming increasingly clear that these tumors are clinically and biologically heterogeneous [19–25].

During the past few years, the Cancer Genome Atlas (TCGA) project has exposed the complexity of the genome-wide genetic alterations in breast cancer [26]. On the other hand, the proper clinical management of Luminal (i.e., ER+) breast cancers, particularly in intermediate-risk patients, remains a matter of controversy. However, there is a limited understanding of how the mutational landscape of these tumors, according to the PR status, can be exploited in the clinic to allow for more tailored management schemes. In this study, we sought: (i) to characterize the mutational signatures of ER+/PR− breast cancers; (ii) to compare the molecular landscapes of PR− and PR+ Luminal tumors; and (iii) to define the prognostic value of the type and pattern of somatic genetic alterations in these patients.

2. Results

A total of 3589 ER+ breast cancers from the publicly available datasets TCGA, Memorial Sloan Kettering (MSK), and Molecular Taxonomy of Breast Cancer International Consortium (METABRIC) were identified. Among them, 3570 (99.5%) cases (2815 invasive ductal carcinomas and 755 invasive carcinomas of any special type) had information on PR status (PR− n = 959, 27%; PR+ n = 2611, 73%) and were included in the current study. The median age at diagnosis of PR− tumors was 59 years old (range 24–92); for PR+ tumors, it was 57 years old (range 23–91). Taken together, 53,585 mutations targeting 13,402 genes were identified, including 57,448 (99%), 6642 (90%), and 8905 (89%) mutations that were private to only one sample in the TCGA, MSK, and METABRIC cohorts, respectively. The number of samples, mutated genes, and mutations of the tumors included in the analysis are summarized in Table 1 and Table S1.

Table 1. Number ER+ breast cancer samples, according to the PR status from the TCGA, MSK, and METABRIC projects. PR, progesterone receptor.

	TCGA (%)	MSK (%)	METABRIC (%)
PR− (n = 959)	110 (12)	396 (41)	453 (47)
PR+ (n = 2611)	608 (23)	1031 (40)	972 (37)
Total (n = 3570)	718 (20)	1427 (40)	1425 (40)

2.1. The Molecular Landscape of ER+/PR− Breast Cancers

The average number of mutations displayed by ER+/PR− breast cancers was 16 per sample, whereas in PR+ tumors was 14. The two groups shared 5668 mutated genes, while approximately 1319 (19%) genes were found to be privately altered in ER+/PR− breast cancers. Overall, the mutations in PR− tumors were missense in 12,583 (78%), nonsense in 1250 (8%), frameshift deletions in 896 (5%), frameshift insertions in 616 (4%), splicing in 516 (3%), and in-frame indels in 261 (2%) cases. Of note, fusion genes were detected in 69 ER+/PR− tumors. The mutational landscape and selected clinicopathologic features in ER+/PR− and ER+/PR+ breast cancers are depicted in Figure 1 and Figure S1, respectively.

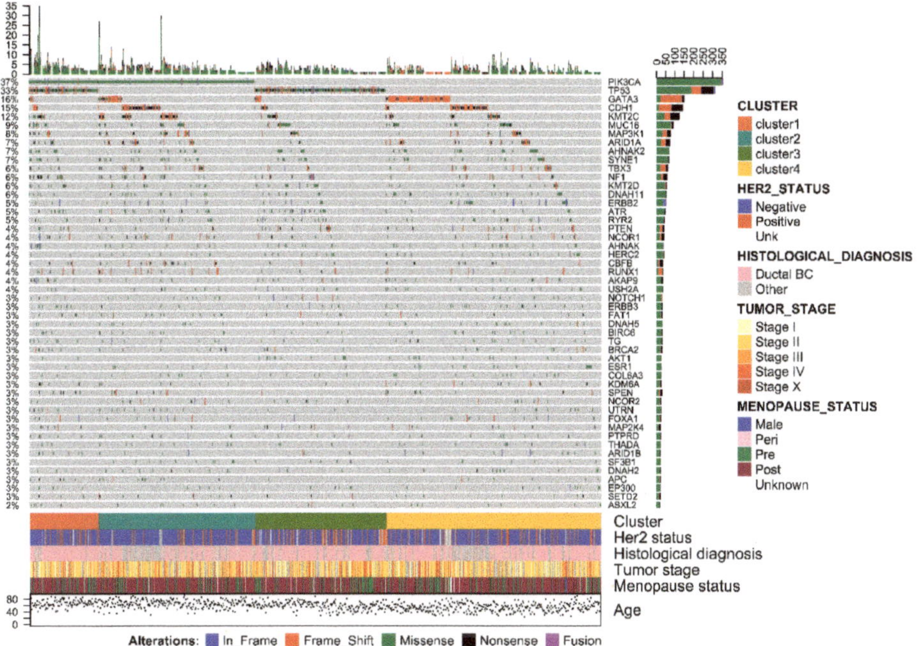

Figure 1. Oncoprint visualization of highly recurrent somatic molecular alterations in ER+/PR− breast cancers (959 samples). Each row represents a gene, as reported on the right, and was sorted by gene alterations frequency (bar plot on the right); types of alterations are color-coded on the basis of the legend on the bottom. Each column represents a sample and was sorted to appreciate the mutual exclusivity across genes. The bar plot on the top represents the number of samples showing alterations in the displayed genes. Cluster analysis, human epidermal growth factor receptor (HER)2 status, histological type, tumor stage, menopause status, and age at diagnosis are reported as rows at the bottom of the figure. Clustering was performed according to the mutual exclusivity and patterns of mutations.

The most frequently mutated gene in PR− tumors was phosphatidylinositol-4,5-bisphosphate 3-kinase, catalytic subunit alpha (*PIK3CA*), with lower prevalence than in PR+ tumors ($n = 354$, 37% vs. $n = 1220$, 47%; $p < 0.01$). In particular, the vast majority of *PIK3CA* mutations were missense and affected four hotspot regions of the gene, namely N345K, E542K, E545K, and H1047R (Figure 2). Notably, the H1047R and E545K mutations in *PIK3CA* were less frequent in PR− tumors (Table 2). The prevalence of samples showing mutations in *TP53*, which was the second most frequently mutated gene in both PR− and PR+ Luminal tumors, was higher in PR− breast cancers ($n = 312$, 33% vs. $n = 496$, 19%; $p < 0.01$). Furthermore, the nonsense mutation R342X and the missense mutations P728S, I195T, and H179R in *TP53* were enriched in PR− tumors ($p < 0.05$), as shown in Table 2. Taken together, *PIK3CA* and *TP53* status allowed for the definition of four molecular clusters (Figure 1). Specifically, Cluster 1 included all *PIK3CA*-mutant/*TP53*-mutant samples ($n = 108$, 11%), Cluster 2 all *PIK3CA*-mutant/*TP53* wild-type samples ($n = 246$, 26%), Cluster 3 *PIK3CA* wild-type/*TP53*-mutant tumors ($n = 204$, 21%), and Cluster

4 encompassed all *PIK3CA/TP53* wild-type cases (n = 401, 42%). Among the other recurrent gene alterations, the hotspot mutation E17K in RAC-alpha serine/threonine-protein kinase (*AKT1*), which was present in 3% and 5% of PR− and PR+ cases, respectively, was mutually exclusive with mutations targeting *PIK3CA*, regardless of PR status (Figure S2). On the other hand, even if *PIK3CA* and *AKT1* were observed to be recurrently mutated in both groups, the hotspot regions differed significantly on the basis of PR activation (p < 0.05). Of note, *GATA3* showed a high number of frame-shift indels and nonsense mutations (Figure 2), consistent with its crucial role in the ER signaling pathway. One of the most recurrently mutated genes was E-cadherin (*CDH1*), with the hotspot truncating mutation in position 23 associated to the lobular histology (Figure 2). The prevalence of human epidermal growth factor receptor (HER)2-mutant cases was higher in PR− breast cancers, albeit nonsignificant (n = 151, 16% vs. n = 389, 15%; p = 0.530). According to the Student's *t*-test, the mutational profile of PR− Luminal breast cancers was significantly different to that of PR+ tumors ($p < 10^{-5}$), with 16 mutations being restricted to the ER+/PR− group, including mutations in *ARID1A, ATR, BCL6, BRAF, CARD11, CDH1, AXIN2, GATA3, MUC16, CCDC82, RUNX1*, and *TBX3* (Table 2). No significant correlations were observed between PR activation status and other clinicopathologic characteristics. The tumor mutational burden (median of five mutations per sample for both PR+/−; mean 15.2 per sample for PR+; mean 15.9 per sample for PR−; range 1–3474 in PR+; and range 1–2900 in PR−) of the cases included in this study is shown in Figure S3.

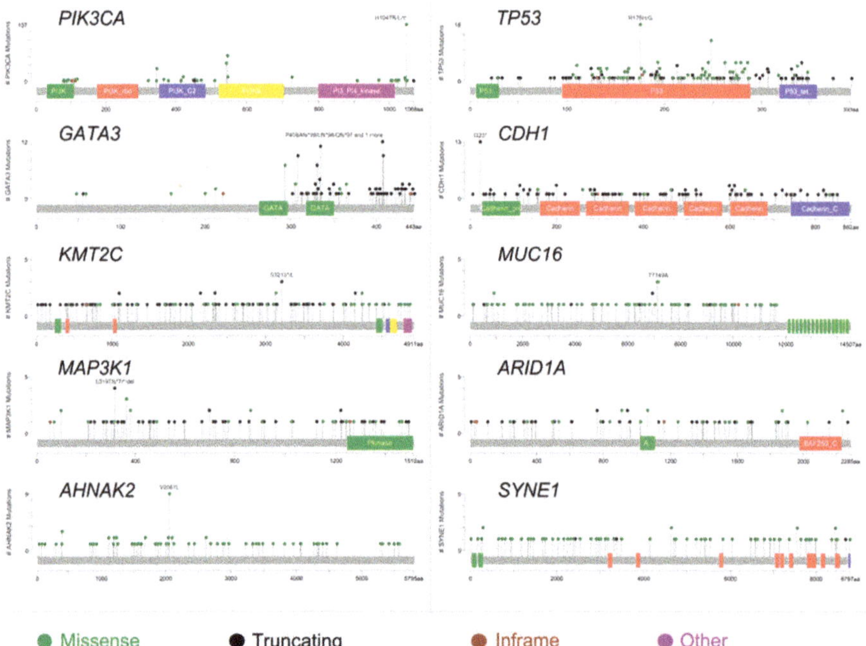

Figure 2. Type of mutations and affected protein domains of the 10 most frequently altered genes in ER+/PR− breast cancers. Mutation types are color-coded on the basis of the legend at the bottom.

Table 2. The 37 recurrent mutations showing significant differences between ER+/PR− and ER+/PR+ breast cancers according to the test of Equal or Given Proportions.

Mutation	PR+ (%)	PR− (%)	*p* Value
ARID1A_Q766SfsX67	0	2 (0.20)	0.019
ATR_A14S	0	2 (0.20)	0.019
BCL6_K474EfsX26	0	2 (0.20)	0.019
BRAF_V600E	0	2 (0.20)	0.019
CARD11_D200E	0	2 (0.20)	0.019
CDH1_R598X	0	2 (0.20)	0.019
CDH1_E138X	0	2 (0.20)	0.019
CDH1_E497RfsX25	0	2 (0.20)	0.019
AXIN2_S493L	0	3 (0.31)	0.004
GATA3_R364T	0	3 (0.31)	0.005
CDH1_V202CfsX7	0	3 (0.31)	0.006
MUC16_T7149A	0	3 (0.31)	0.007
CCDC82_E175del	0	3 (0.31)	0.008
RUNX1_D123GfsX15	0	4 (0.41)	<0.001
TBX3_W113X	0	4 (0.41)	<0.001
CDH1_T115NfsX53	1 (0.04)	3 (0.31)	0.029
FOXA1_D226N	1 (0.04)	3 (0.31)	0.029
FOXA1_I176M	1 (0.04)	3 (0.31)	0.029
GATA3_X444LfsX63	1 (0.04)	3 (0.31)	0.029
TERT_Promoter	1 (0.04)	3 (0.31)	0.029
TP53_P278S	1 (0.04)	3 (0.31)	0.029
SMAD4_Q245X	1 (0.04)	3 (0.31)	0.029
TP53_I195T	1 (0.04)	5 (0.52)	0.002
ERBB2_E770_A771insGIRD	1 (0.04)	8 (0.83)	0.003
ERBB2_S310F	2 (0.08)	4 (0.41)	0.027
MAP3K1_R364W	2 (0.08)	4 (0.41)	0.027
TP53_H179R	2 (0.08)	4 (0.41)	0.027
TP53_R342X	5 (0.19)	7 (0.72)	0.013
GATA3_D335GfsX17	16 (0.61)	13 (1.35)	0.028
TP53_R175H	21 (0.80)	18 (1.87)	0.006
ESR1_Y537S	29 (1.11)	3 (0.31)	0.024
ESR1_D538G	47 (1.80)	7 (0.72)	0.020
SF3B1_K700E	60 (2.29)	10 (1.04)	0.016
GATA3_X308_splice	70 (2.68)	9 (0.94)	0.002
AKT1_E17K	106 (4.05)	25 (2.60)	0.04
PIK3CA_E545K	251 (9.61)	68 (7.09)	0.019
PIK3CA_H1047R	482 (18.46)	134 (13.97)	0.002

2.2. The Prognostic Role of PIK3CA and TP53 in ER+/PR− Breast Cancers

Overall, the highest mortality was observed before 50 months from the diagnosis, regardless of PR status, with a median survival of 76.9 months in PR− and 61 months in PR+ tumors. The most recurrently mutated genes in ER+/PR− and ER+/PR+ breast cancers were used to define the survival probability curves shown in Figures S4 and S5, respectively. Even though the log-rank *p*-values were significant for *TP53* and *GATA3* mutations in both groups, survival analyses including tumors harboring alterations only in each of the most frequently mutated genes, but not in the others, revealed that in ER+/PR− breast cancers only *TP53* mutations are related to a different prognosis (Figure 3). The hotspot regions of *TP53* that were significantly different in PR− tumors were not related to a different outcome (Figure S6), similar to *PIK3CA* (Figure S7).

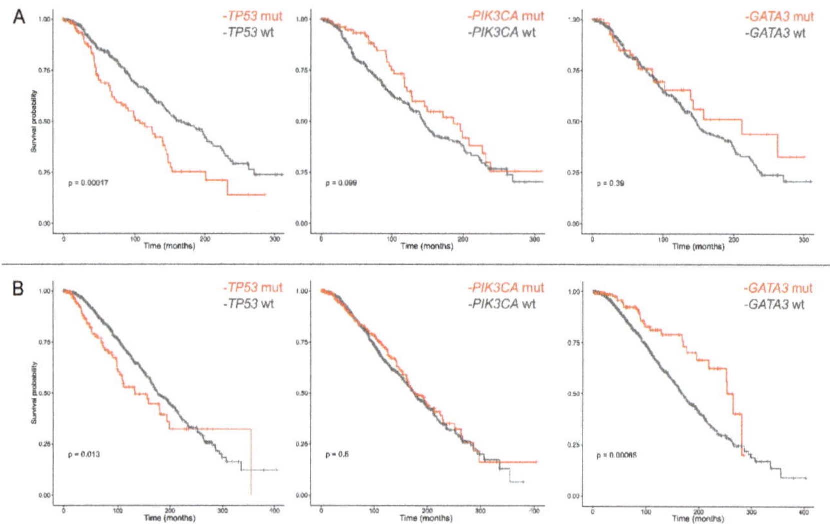

Figure 3. Overall survival of ER+/PR− (**A**) and ER+/PR+ (**B**) breast cancer patients based on *TP53*, *PIK3CA*, and *GATA3* gene alterations. For each analysis, all samples harboring mutations in one of the other two genes were excluded. Survival curves are built according to the Kaplan–Meier method.

Subsequently, survival curves were built according to the four molecular clusters identified on the basis of *PIK3CA* and *TP53* status. These analyses revealed the prognostic value of the combination and mutual exclusivity of *PIK3CA* and *TP53* mutations (Figure 4). Specifically, Cluster 4 showed in both PR− and PR+ cases a good prognosis. Interestingly, the prognosis of Cluster 4 overlapped to that of Cluster 3 in PR+ but not in PR− tumors. Hence, PR− breast cancers showed a different scenario, where the long-term outcome of the patients was worse in the presence of *PIK3CA* and/or *TP53* mutations (i.e., Clusters 1, 2 and 3).

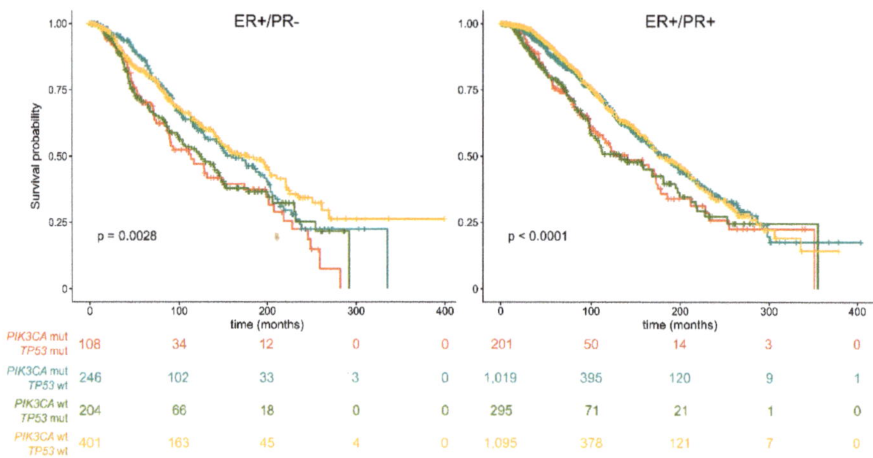

Figure 4. Overall survival of *PIK3CA*/*TP53*-mutant (Cluster 1), *PIK3CA*mutant/*TP53* wild-type (Cluster 2), *PIK3CA* wild-type/*TP53*-mutant (Cluster 3), and *PIK3CA*/*TP53* wild-type (Cluster 4) ER+ breast cancers, based on PR activation. Survival curves are built according to the Kaplan–Meier method.

3. Discussion

The precise risk stratification in Luminal breast cancer by means of immunohistochemistry and/or prognostic genomic tests is a major limitation in defining the most appropriate management scheme [27]. Patients with ER+ breast cancers are assumed to have a good prognosis, but the lack of PR expression may contribute to their poor outcomes. This may be a result of the de-differentiation of hormone-positive neoplasms and subsequent development of resistance phenomena to both anti-estrogen therapy and chemotherapy. Studies aiming to explore the genetic alterations in ER+/PR− breast cancers have been performed. However, the unique biology and challenging clinical course of these tumors, particularly in long-term survivors, suggest that they warrant further characterization. In this study, we analyzed a large cohort of PR− Luminal breast cancers with publicly available genomic data and compared their molecular landscape and prognosis to those of PR+ tumors. Altogether, we observed that several alterations in clinically actionable cancer genes are private to or enriched for in PR− breast cancers, such as *TP53* R342X, P728S, I195T, and H179R, *GATA3*, *CDH1*, *HER2*, *CDH1*, and *BRAF* V600E. Furthermore, we identified four molecular clusters on the basis of *PIK3CA* and *TP53* status with significantly different risk of death in PR− tumors.

Decreased expression and/or downregulation of PR in breast cancer leads to a subset of tumors that is phenotypically ER+/PR−. Even though several hypotheses to explain this phenomenon have been put forward, we are still far from fully understanding its biology. In a proportion of Luminal tumors, ER, although expressed, is biologically nonfunctional and therefore it is unable to stimulate PR production, particularly in postmenopausal women [1]. Another mechanism for PR loss is the epigenetic inactivation of its promoter through hypermethylation [12]. Even though a genetic loss of a PR gene locus has previously been observed [12], in our analysis, all ER+/PR− tumors are PR wild-type, suggesting that PR downregulation may be determined by growth factor pathways, as previously observed [2,11]. In particular, the HER2 activity may lead to the cytoplasmic sequestration of ER, which alters a set of genes that are normally regulated by ER, including PR−related genes, such as *PIK3CA* [11,28,29].

Taken together, we observed that the most frequently mutated genes in ER+/PR− breast cancers are *PIK3CA*, *TP53*, *GATA3*, *CHD1*, *KMT2C*, *MUC16*, *MAP3K1*, *ARID1A*, *AHNAK2*, and *SYNE2*. Interestingly, *PIK3CA* and *TP53* show a mutational prevalence (37% and 33%, respectively) that differs significantly to that of ER+/PR+ tumors (with *PIK3CA* mutated in 47% of cases and *TP53* in 19%). Those aspects have already been described in the literature [30,31]. On the other hand, the identification of a mutational profile specific to ER+/PR− cases, with 16 mutations being restricted to this group, is a novel finding. In our study, we confirm the presence of highly recurrent molecular alterations of the *PIK3CA* gene in position 1047, which likely constitute the driving genetic event in the pathogenesis of a subset of ER+/PR− breast cancers. These data provide further credence to the notion that inhibitors of this pathway (e.g., XL147) could reverse PR downregulation and overcome resistance to anti-HER2 drugs [32]. In addition, the identification of the *BRAF* V600E as a private mutation of PR− cases have possible therapeutic implications [33,34]. Recently, mutations in *HER2* have been detected in breast cancer patient samples which lack *HER2* gene amplification. Thirteen *HER2* mutations were characterized from twenty-five patient samples which had HER2 mutations but lacked HER2 gene amplification. Among them, seven mutations were activating and resulted from point mutations and in-frame deletions. Some mutations (L755S) resulted in lapatinib resistance; however, this was not an activating mutation. All of the cells containing the *HER2* mutations were sensitive to the irreversible HER2 kinase inhibitor, neratinib [35]. Our analysis corroborates the concept that mutations in *GATA3* are associated with a better outcome in ER+ breast cancer patients [36]. After eliminating all cases with concurrent mutations in the other top recurrently mutated genes, however, we were able to confirm this notion only in PR+ tumors. These data suggest that *GATA3* mutations are not an independent good prognostic factor in ER+/PR− tumors. Given that *GATA3* is frequently altered in Luminal A breast cancers, our findings provide an additional molecular layer to the worse prognosis showed by ER+/PR− breast cancers [19,37]. Furthermore, we confirmed that *TP53* mutations are associated with

PR negativity and with a shorter overall survival time in breast cancers [38]. Interestingly, this behavior is unrelated to the specific regions of *TP53* that are recurrently altered in this subset of patients, akin to patients with *PIK3CA*-mutant tumors.

The patterns of mutations in *TP53* with *PIK3CA* allowed us to identify four molecular clusters in both PR− and PR+ Luminal breast cancers, namely *PIK3CA/TP53*-mutant (Cluster 1), *PIK3CA*-mutant/*TP53* wild-type (Cluster 2), *PIK3CA* wild-type/*TP53*-mutant (Cluster 3), and *PIK3CA/TP53* wild-type (Cluster 4). Notably, the prognostic distribution of these clusters differed substantially between ER+/PR− and ER+/PR+ breast cancers. Indeed, while in PR+ Luminal tumors Clusters 2 and 4 were related to better survival, with overlapping curves, in PR− Luminal tumors Cluster 2 followed into in an intermediate risk category for the first 16 years of follow-up, becoming worse after that time. All these diverse correlations highlight the importance of *PIK3CA* and *TP53* analysis in PR− Luminal breast cancer prognostication.

4. Materials and Methods

4.1. Case Selection and Definitions

We used the CGDS R package to interrogate the Cancer Genomics Data Server [39,40] and download mutational and clinical data related to three breast cancer projects hosted at the Memorial-Sloan-Kettering Cancer Center: the METABRIC project [41], containing 2509 breast cancers samples; the MSK project [42] containing 1918 samples; and the TCGA project, containing 1105 samples. Each sample has both somatic mutational profiles for selected genes, and clinical information. In particular, the TCGA project contains mutational profiles for 20,461 genes, the METABTIC project contains mutational profiles for 173 genes and the MSK project for 474 genes. Moreover, we used gplots and ggplot2 packages [43,44] to perform the clustering analysis and visualize the data. We collected all somatic mutations related to the projects and integrated them to the clinical information and the treatment outcomes. Moreover, we selected all the estrogen receptor positive (ER+) samples reducing our dataset to 3589 samples, and a total of 53,585 somatic mutations in 13,402 genes.

4.2. Statistical Analysis

Comparisons between groups were generally performed using the Student's *t*-test and test of Equal or Given Proportions. Event-free survival was expressed as the number of months from diagnosis to the occurrence of distant or local relapse or death (disease-related death). Cumulative survival probabilities were calculated using the Kaplan–Meier method. Differences between survival rates were tested with the log-rank test (SPSS version 20.0; IBM). Survival data were censored at five years. A $p < 0.05$ was considered statistically significant. Survival analysis and figures were developed using the R survival and survminer packages [45], and the Kaplan–Meier non-parametric statistic [46].

5. Conclusions

We demonstrated that ER+/PR− breast cancers are biologically characterized by relevant molecular characteristics in terms of prognostic and predictive information, which could be integrated into the clinical setting to realize the potentials of precision medicine in these clinically, and pathologically, challenging neoplasms.

Supplementary Materials: Supplementary materials can be found at http://www.mdpi.com/1422-0067/20/3/510/s1. Figure S1: Oncoprint visualization of highly recurrent somatic molecular alterations in ER+/PR+ breast cancers (2611 samples). Each row represents a gene, as reported on the right and was sorted by gene alterations frequency (bar plot on the right); types of alterations are color-coded on the basis of the legend on the bottom. Each column represents a sample and was sorted to appreciate the mutual exclusivity across genes. The bar plot on the top represents the number of samples showing alterations in the displayed genes. Cluster analysis, HER2 status, histological type, tumor stage, and menopause status are reported as rows at the bottom of the figure; age at diagnosis is depicted in the top at the bottom. Clustering was performed according to the mutual exclusivity and patterns of mutations. Figure S2: Recurrent somatic alterations in 959 ER+/PR− (A) and in 2611 ER+/PR+ (B) breast cancers (2611 samples). Each row represents an alteration, as reported on the right, each column a sample. Alterations were sorted by frequency, while the samples were sorted to appreciate the mutual exclusivity across alterations. Figures report the 50 most frequent gene alterations. Figure S3: Total number of mutations per samples in ER+/PR− (**A**) and ER+/PR+ (**B**) breast cancer patients. Each bar represents a sample; types of alterations are color-coded on the basis of the legend on the left. Figure S4: Overall survival of ER+/PR− breast cancer patients based on the most frequently altered genes. Survival curves (red, mutant; gray, wild-type) are built according to the Kaplan–Meier method. For each analysis, all samples harboring mutations in one of the other nine genes were excluded. Figure S5: Overall survival of ER+/PR+ breast cancer patients based on the most frequently altered genes. For each analysis, all samples harboring mutations in one of the other nine genes were excluded. Survival curves (red, mutant; gray, wild-type) are built according to the Kaplan–Meier method. Figure S6: Overall survival of ER+/PR− breast cancer patients based on the most frequently altered regions in the *PIK3CA* gene. Survival curves are built according to the Kaplan–Meier method. Figure S7: Overall survival of ER+/PR− breast cancer patients based on the most frequently altered regions in the *TP53* gene. Survival curves are built according to the Kaplan–Meier method. Table S1: Mutations of the tumors included in the analysis.

Author Contributions: Conceptualization, N.F.; Methodology, G.L., J.C. and N.F.; Software, J.C.; Formal Analysis, J.C. and N.F.; Investigation, G.L., M.C., L.F. and N.F.; Resources, S.F. and M.M.; Data Curation, J.C. and N.F.; Writing—Original Draft Preparation, G.L., J.C. and M.C.; Writing—Review and Editing, L.F., S.F., M.M. and N.F.; and Supervision, S.F., M.M. and N.F.

Funding: This research received no external funding.

Acknowledgments: The results of this study are based on data generated by The Cancer Genome Atlas (TCGA) Research Network: http://cancergenome.nih.gov/, which the authors would like to thank for making available all data without restrictions on their use in publications.

References

1. Thakkar, J.P.; Mehta, D.G. A review of an unfavorable subset of breast cancer: Estrogen receptor positive progesterone receptor negative. *Oncologist* **2011**, *16*, 276–285. [CrossRef] [PubMed]

2. Arpino, G.; Weiss, H.; Lee, A.V.; Schiff, R.; De Placido, S.; Osborne, C.K.; Elledge, R.M. Estrogen receptor-positive, progesterone receptor-negative breast cancer: Association with growth factor receptor expression and tamoxifen resistance. *J. Natl. Cancer Inst.* **2005**, *97*, 1254–1261. [CrossRef] [PubMed]

3. *WHO Classification of Tumours of the Breast*; International Agency for Research on Cancer: Lyon, France, 2012.

4. Purdie, C.A.; Quinlan, P.; Jordan, L.B.; Ashfield, A.; Ogston, S.; Dewar, J.A.; Thompson, A.M. Progesterone receptor expression is an independent prognostic variable in early breast cancer: A population-based study. *Br. J. Cancer* **2014**, *110*, 565–572. [CrossRef] [PubMed]

5. Van Mackelenbergh, M.T.; Denkert, C.; Nekljudova, V.; Karn, T.; Schem, C.; Marme, F.; Stickeler, E.; Jackisch, C.; Hanusch, C.; Huober, J.; et al. Outcome after neoadjuvant chemotherapy in estrogen receptor-positive and progesterone receptor-negative breast cancer patients: A pooled analysis of individual patient data from ten prospectively randomized controlled neoadjuvant trials. *Breast Cancer Res. Treat.* **2018**, *167*, 59–71. [CrossRef] [PubMed]

6. Yu, K.D.; Liu, G.Y.; Di, G.H.; Wu, J.; Lu, J.S.; Shen, K.W.; Shen, Z.Z.; Shao, Z.M. Progesterone receptor status provides predictive value for adjuvant endocrine therapy in older estrogen receptor-positive breast cancer patients. *Breast* **2007**, *16*, 307–315. [CrossRef] [PubMed]

7. Neven, P.; Pochet, N.; Drijkoningen, M.; Amant, F.; De Smet, F.; Paridaens, R.; Christiaens, M.R.; Vergote, I. Progesterone receptor in estrogen receptor-positive breast cancer: The association between her-2 and lymph node involvement is age related. *J. Clin. Oncol.* **2006**, *24*, 2595–2597. [CrossRef] [PubMed]

8. Viale, G.; Regan, M.M.; Maiorano, E.; Mastropasqua, M.G.; Dell'Orto, P.; Rasmussen, B.B.; Raffoul, J.; Neven, P.; Orosz, Z.; Braye, S.; et al. Prognostic and predictive value of centrally reviewed expression of estrogen and progesterone receptors in a randomized trial comparing letrozole and tamoxifen adjuvant therapy for postmenopausal early breast cancer: Big 1-98. *J. Clin. Oncol.* **2007**, *25*, 3846–3852. [CrossRef] [PubMed]

9. Olivotto, I.A.; Truong, P.T.; Speers, C.H.; Bernstein, V.; Allan, S.J.; Kelly, S.J.; Lesperance, M.L. Time to stop progesterone receptor testing in breast cancer management. *J. Clin. Oncol.* **2004**, *22*, 1769–1770. [CrossRef] [PubMed]

10. Burstein, H.J.; Prestrud, A.A.; Seidenfeld, J.; Anderson, H.; Buchholz, T.A.; Davidson, N.E.; Gelmon, K.E.; Giordano, S.H.; Hudis, C.A.; Malin, J.; et al. American society of clinical oncology clinical practice guideline: Update on adjuvant endocrine therapy for women with hormone receptor-positive breast cancer. *J. Clin. Oncol.* **2010**, *28*, 3784–3796. [CrossRef] [PubMed]

11. Osborne, C.K.; Shou, J.; Massarweh, S.; Schiff, R. Crosstalk between estrogen receptor and growth factor receptor pathways as a cause for endocrine therapy resistance in breast cancer. *Clin. Cancer Res.* **2005**, *11*, 865s–870s.

12. Cui, X.; Schiff, R.; Arpino, G.; Osborne, C.K.; Lee, A.V. Biology of progesterone receptor loss in breast cancer and its implications for endocrine therapy. *J. Clin. Oncol.* **2005**, *23*, 7721–7735. [CrossRef] [PubMed]

13. Creighton, C.J.; Kent Osborne, C.; van de Vijver, M.J.; Foekens, J.A.; Klijn, J.G.; Horlings, H.M.; Nuyten, D.; Wang, Y.; Zhang, Y.; Chamness, G.C.; et al. Molecular profiles of progesterone receptor loss in human breast tumors. *Breast Cancer Res. Treat.* **2009**, *114*, 287–299. [CrossRef] [PubMed]

14. Maggi, A. Liganded and unliganded activation of estrogen receptor and hormone replacement therapies. *Biochim. Biophys. Acta* **2011**, *1812*, 1054–1060. [CrossRef] [PubMed]

15. Li, W.; Jia, M.; Qin, X.; Hu, J.; Zhang, X.; Zhou, G. Harmful effect of erbeta on bcrp-mediated drug resistance and cell proliferation in eralpha/PR−negative breast cancer. *FEBS J.* **2013**, *280*, 6128–6140. [CrossRef] [PubMed]

16. Clarke, R.; Tyson, J.J.; Dixon, J.M. Endocrine resistance in breast cancer–an overview and update. *Mol. Cell Endocrinol.* **2015**, *418 Pt 3*, 220–234. [CrossRef]

17. Johnston, S.R.; Saccani-Jotti, G.; Smith, I.E.; Salter, J.; Newby, J.; Coppen, M.; Ebbs, S.R.; Dowsett, M. Changes in estrogen receptor, progesterone receptor, and ps2 expression in tamoxifen-resistant human breast cancer. *Cancer Res.* **1995**, *55*, 3331–3338. [PubMed]

18. Thomas, C.; Gustafsson, J.A. Progesterone receptor-estrogen receptor crosstalk: A novel insight. *Trends Endocrinol. Metab.* **2015**, *26*, 453–454. [CrossRef] [PubMed]

19. Piscuoglio, S.; Ng, C.K.; Murray, M.P.; Guerini-Rocco, E.; Martelotto, L.G.; Geyer, F.C.; Bidard, F.C.; Berman, S.; Fusco, N.; Sakr, R.A.; et al. The genomic landscape of male breast cancers. *Clin. Cancer Res.* **2016**. [CrossRef] [PubMed]

20. Fusco, N.; Geyer, F.C.; De Filippo, M.R.; Martelotto, L.G.; Ng, C.K.; Piscuoglio, S.; Guerini-Rocco, E.; Schultheis, A.M.; Fuhrmann, L.; Wang, L.; et al. Genetic events in the progression of adenoid cystic carcinoma of the breast to high-grade triple-negative breast cancer. *Mod. Pathol.* **2016**, *29*, 1292–1305. [CrossRef]

21. Kim, J.; Geyer, F.C.; Martelotto, L.G.; Ng, C.K.Y.; Lim, R.S.; Selenica, P.; Li, A.; Pareja, F.; Fusco, N.; Edelweiss, M.; et al. Mybl1 rearrangements and myb amplification in breast adenoid cystic carcinomas lacking the myb-nfib fusion gene. *J. Pathol.* **2018**, *244*, 143–150. [CrossRef]

22. Fusco, N.; Colombo, P.E.; Martelotto, L.G.; De Filippo, M.R.; Piscuoglio, S.; Ng, C.K.; Lim, R.S.; Jacot, W.; Vincent-Salomon, A.; Reis-Filho, J.S.; et al. Resolving quandaries: Basaloid adenoid cystic carcinoma or breast cylindroma? The role of massively parallel sequencing. *Histopathology* **2016**, *68*, 262–271. [CrossRef] [PubMed]

23. Marchiò, C.; De Filippo, M.R.; Ng, C.K.; Piscuoglio, S.; Soslow, R.A.; Reis-Filho, J.S.; Weigelt, B. Piking the type and pattern of pi3k pathway mutations in endometrioid endometrial carcinomas. *Gynecol. Oncol.* **2015**, *137*, 321–328. [CrossRef] [PubMed]

24. De Mattos-Arruda, L.; Bidard, F.C.; Won, H.H.; Cortes, J.; Ng, C.K.; Peg, V.; Nuciforo, P.; Jungbluth, A.A.; Weigelt, B.; Berger, M.F.; et al. Establishing the origin of metastatic deposits in the setting of multiple primary malignancies: The role of massively parallel sequencing. *Mol. Oncol.* **2014**, *8*, 150–158. [CrossRef]

25. Ng, C.K.; Pemberton, H.N.; Reis-Filho, J.S. Breast cancer intratumor genetic heterogeneity: Causes and implications. *Expert. Rev. Anticancer Ther.* **2012**, *12*, 1021–1032. [CrossRef] [PubMed]

26. Kandoth, C.; McLellan, M.D.; Vandin, F.; Ye, K.; Niu, B.; Lu, C.; Xie, M.; Zhang, Q.; McMichael, J.F.; Wyczalkowski, M.A.; et al. Mutational landscape and significance across 12 major cancer types. *Nature* **2013**, *502*, 333–339. [CrossRef] [PubMed]

27. Fusco, N.; Lopez, G.; Corti, C.; Pesenti, C.; Colapietro, P.; Ercoli, G.; Gaudioso, G.; Faversani, A.; Gambini, D.; Michelotti, A.; et al. Mismatch repair protein loss as a prognostic and predictive biomarker in breast cancers regardless of microsatellite instability. *JNCI Cancer Spectrum* **2018**, *2*, pky056. [CrossRef]

28. Ercoli, G.; Lopez, G.; Ciapponi, C.; Corti, C.; Despini, L.; Gambini, D.; Runza, L.; Blundo, C.; Sciarra, A.; Fusco, N. Building up a high-throughput screening platform to assess the heterogeneity of HER2 gene amplification in breast cancers. *J. Vis. Exp.* **2017**, *13*, 233–236. [CrossRef]

29. Fusco, N.; Bosari, S. HER2 aberrations and heterogeneity in cancers of the digestive system: Implications for pathologists and gastroenterologists. *World J. Gastroenterol.* **2016**, *22*, 7926–7937. [CrossRef]

30. Vuong, D.; Simpson, P.T.; Green, B.; Cummings, M.C.; Lakhani, S.R. Molecular classification of breast cancer. *Virchows Arch.* **2014**, *465*, 1–14. [CrossRef]

31. Russnes, H.G.; Lingjaerde, O.C.; Borresen-Dale, A.L.; Caldas, C. Breast cancer molecular stratification: From intrinsic subtypes to integrative clusters. *Am. J. Pathol.* **2017**, *187*, 2152–2162. [CrossRef]

32. Chakrabarty, A.; Bhola, N.E.; Sutton, C.; Ghosh, R.; Kuba, M.G.; Dave, B.; Chang, J.C.; Arteaga, C.L. Trastuzumab-resistant cells rely on a her2-pi3k-foxo-survivin axis and are sensitive to pi3k inhibitors. *Cancer Res.* **2013**, *73*, 1190–1200. [CrossRef] [PubMed]

33. Garbe, C.; Abusaif, S.; Eigentler, T.K. Vemurafenib. *Recent Results Cancer Res.* **2014**, *201*, 215–225. [PubMed]

34. Fumagalli, C.; Bianchi, F.; Raviele, P.R.; Vacirca, D.; Bertalot, G.; Rampinelli, C.; Lazzeroni, M.; Bonanni, B.; Veronesi, G.; Fusco, N.; et al. Circulating and tissue biomarkers in early-stage non-small cell lung cancer. *Ecancermedicalscience* **2017**, *11*, 717. [CrossRef] [PubMed]

35. Fda Approved Drug: Neratinib. Available online: https://www.accessdata.fda.gov/drugsatfda_docs/label/2017/208051s000lbl.pdf (accessed on 15 December 2018).

36. Miettinen, M.; McCue, P.A.; Sarlomo-Rikala, M.; Rys, J.; Czapiewski, P.; Wazny, K.; Langfort, R.; Waloszczyk, P.; Biernat, W.; Lasota, J.; et al. Gata3: A multispecific but potentially useful marker in surgical pathology: A systematic analysis of 2500 epithelial and nonepithelial tumors. *Am. J. Surg. Pathol.* **2014**, *38*, 13–22. [CrossRef] [PubMed]

37. Ng, C.K.; Schultheis, A.M.; Bidard, F.C.; Weigelt, B.; Reis-Filho, J.S. Breast cancer genomics from microarrays to massively parallel sequencing: Paradigms and new insights. *J. Natl. Cancer Inst.* **2015**, *107*. [CrossRef] [PubMed]

38. Silwal-Pandit, L.; Vollan, H.K.; Chin, S.F.; Rueda, O.M.; McKinney, S.; Osako, T.; Quigley, D.A.; Kristensen, V.N.; Aparicio, S.; Borresen-Dale, A.L.; et al. Tp53 mutation spectrum in breast cancer is subtype specific and has distinct prognostic relevance. *Clin. Cancer Res.* **2014**, *20*, 3569–3580. [CrossRef] [PubMed]

39. Cran—Package Cgdsr. Available online: https://cran.r-project.org/web/packages/cgdsr/index.html (accessed on 15 December 2018).

40. R: The R Project for Statistical Computing. Available online: https://www.r-project.org/ (accessed on 15 December 2018).

41. Pereira, B.; Chin, S.F.; Rueda, O.M.; Vollan, H.K.; Provenzano, E.; Bardwell, H.A.; Pugh, M.; Jones, L.; Russell, R.; Sammut, S.J.; et al. The somatic mutation profiles of 2,433 breast cancers refines their genomic and transcriptomic landscapes. *Nat. Commun.* **2016**, *7*, 11479. [CrossRef]

42. Razavi, P.; Chang, M.T.; Xu, G.; Bandlamudi, C.; Ross, D.S.; Vasan, N.; Cai, Y.; Bielski, C.M.; Donoghue, M.T.A.; Jonsson, P.; et al. The genomic landscape of endocrine-resistant advanced breast cancers. *Cancer Cell* **2018**, *34*, 427–438. [CrossRef]

43. Cran—Package Gplots. Available online: https://cran.r-project.org/web/packages/gplots/index.html (accessed on 15 December 2018).

44. Wickham, H. *Ggplot2—Elegant Graphics for Data Analysis*; Springer: New York, NY, USA, 2018.

45. Cran—Package Survival. Available online: https://cran.r-project.org/web/packages/survival/index.html (accessed on 15 December 2018).

46. Kaplan, E.L.; Meier, P. Nonparametric estimation from incomplete observations. *J. Am. Stat. Assoc.* **1958**, *53*, 457–481. [CrossRef]

International Journal of
Molecular Sciences

Review

G-Protein Coupled Estrogen Receptor in Breast Cancer

Li-Han Hsu [1,2,3], Nei-Min Chu [4], Yung-Feng Lin [1,5] and Shu-Huei Kao [1,5,*]

[1] Ph.D. Program in Medical Biotechnology, College of Medical Science and Technology,
     Taipei Medical University, Taipei 110, Taiwan; lhhsu@kfsyscc.org (L.-H.H.); yflin@tmu.edu.tw (Y.-F.L.)
[2] Division of Pulmonary and Critical Care Medicine, Sun Yat-Sen Cancer Center, Taipei 112, Taiwan
[3] Department of Medicine, National Yang-Ming University Medical School, Taipei 112, Taiwan
[4] Department of Medical Oncology, Sun Yat-Sen Cancer Center, Taipei 112, Taiwan; nmchu@kfsyscc.org
[5] School of Medical Laboratory Science and Biotechnology, College of Medical Science and Technology,
     Taipei Medical University, 250 Wu-Hsing Street, Taipei 110, Taiwan
*   Correspondence: kaosh@tmu.edu.tw; Tel.: +886-2-2736-1661 (ext. 3317); Fax: +886-2-2732-4510

Received: 29 November 2018; Accepted: 12 January 2019; Published: 14 January 2019

Abstract: The G-protein coupled estrogen receptor (GPER), an alternate estrogen receptor (ER) with a structure distinct from the two canonical ERs, being ERα, and ERβ, is expressed in 50% to 60% of breast cancer tissues and has been presumed to be associated with the development of tamoxifen resistance in ERα positive breast cancer. On the other hand, triple-negative breast cancer (TNBC) constitutes 15% to 20% of breast cancers and frequently displays a more aggressive behavior. GPER is prevalent and involved in TNBC and can be a therapeutic target. However, contradictory results exist regarding the function of GPER in breast cancer, proliferative or pro-apoptotic. A better understanding of the GPER, its role in breast cancer, and the interactions with the ER and epidermal growth factor receptor will be beneficial for the disease management and prevention in the future.

Keywords: breast cancer; epidermal growth factor receptor; estrogen; estrogen receptor; G-protein coupled estrogen receptor

---

1. Introduction

Breast cancer is the most common cancer among women worldwide [1,2]. There were over two million new cases in 2018. In 2015, 14,801 new cases were diagnosed in Taiwan [3]. It ranked fourth in mortality in Taiwan and led to the death of 18.2 persons per 100,000 of the population. Estrogen, predominantly 17β-estradiol (E_2) and its receptor has long been known to enhance the development and progression of breast cancer. Drugs targeting the estrogen signaling pathway through the selective estrogen receptor modulator (SERM) (e.g., tamoxifen, raloxifene), the estrogen receptor (ER) antagonists (e.g., fulvestrant) and, the aromatase inhibitors, including the reversible nonsteroidal agents (e.g., letrozole, anastrozole), or the irreversible steroidal inactivator (e.g., exemestane), has been used for decades to treat ER positive breast cancer (Figure 1) [4]. Tamoxifen is the first SERM approved for the treatment of breast cancer and effectively demonstrated in the reduction of the recurrence and prevention of contralateral breast cancer. However, primary or acquired resistance frequently arises and becomes the major obstacle in hormone therapy, which indicates a more complex receptor and signaling pathways involved in the cancer progression. The G-protein coupled estrogen receptor (GPER), originally known as GPR30, a seven transmembrane domain protein, is an alternate estrogen receptor with a structure distinct from the two canonical estrogen receptors, ERα and ERβ mainly mediate a rapid non-genomic response [5–10]. This is expressed in about 50% to 60% of breast cancer tissues and has been reported as a modulator of neoplastic transformation (Figure 2) [11–25]. Paradoxically, the modulators or antagonists of the classical estrogen receptors such as tamoxifen, raloxifene, and fulvestrant, were found to be the GPER agonists [24]. The expression of GPER has been presumed to be associated with the development of tamoxifen resistance [26–32]. In breast cancer patients treated with tamoxifen, there is an increased risk of developing endometrial cancer and often has a poor clinical outcome. GPER was also supposed to mediate the contrary tissue-specific effect [33,34].

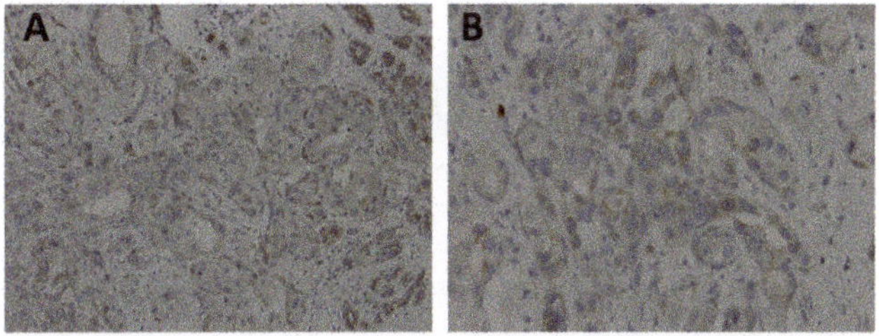

Figure 1. Structures of the representative GPER agonists and antagonists. 17β-estradiol is one of the major physiological forms of estrogen. Tamoxifen is both a selective estrogen receptor modulator and an agonist for the GPER. Bisphenol A is a xenoestrogen. Fulvestrant is a selective estrogen receptor downregulator (ER antagonist) and an agonist for the GPER. G-1 is a selective GPER agonist, whereas G15 is a selective GPER antagonist. Abbreviation: ER, estrogen receptor; GPER, G-protein coupled estrogen receptor.

Figure 2. Representative case of archival, paraffin-embedded breast ductal carcinoma stained with polyclonal GPER1 antibody (Sigma-Aldrich, 1:50 dilution) showed focal, weak membranous and cytoplasmic expression. (**A**) original × 200; and (**B**) original × 400.

On the other hand, triple-negative breast cancer (TNBC), defined as a lack of ERα, progesterone receptor (PR), or the overexpression of human epidermal growth factor receptor 2 (HER2/neu), constitutes 15% to 20% of breast cancers. It is more prevalent in younger women and is frequently present at a more advanced stage with a more aggressive behavior. Lacking a well-defined receptor and signaling pathway, chemotherapy remains the treatment of choice but with a higher rate of recurrence. GPER is prevalent in TNBC and presumed to be involved in the growth of TNBC. It can be considered as a candidate of therapeutic target [35–40].

The endocrine disruptive chemicals, such as bisphenol (Figure 1) and thiodiphenol, at the environmentally relevant doses may exert effects through the GPER and estrogen-like signaling pathways, contribute to breast cancer progression, and drug resistance in both the ERα-positive and -negative breast cancer cells [41–45].

The epidermal growth factor receptor (EGFR) activation is a common and important event in the pathogenesis and progression of breast cancer. The EGFR transactivation by estrogen via the GPER has been proposed as an alternate signaling pathway with a potential significance for breast cancer [46–54].

However, contradictory results have existed regarding the response of GPER to estrogens/antiestrogens and the effect of GPER agonist/antagonist on the proliferation, migration and invasion of the breast cancer cells [55–65]. There were controversies on the subcellular localization of GPER and its function, proliferative or pro-apoptotic. Therefore, the role of GPER in ERα positive breast cancer and TNBC remains unclear.

In the following, updated evidence about the GPER in breast cancer were examined. Future epidemiology and laboratory studies, which may be helpful to elucidate the role of the GPER, were proposed.

2. G-Protein Coupled Estrogen Receptor Expression in Breast Cancer

The significance of GPER in human breast cancer was evaluated by comparing its relationship to ER, PR, and the cancer progression variables through immunohistochemical analysis [15]. A significant association between the GPER and ER was observed. GPER was positively correlated with the HER2/neu expression, tumor size, and metastasis. The distinct patterns of the GPER and ER in association with the cancer progression variable supported that the GPER and ER have independent influences on the estrogen responsiveness of breast cancer. The association between GPER expression and tamoxifen resistance was later confirmed [27]. The GPER was negatively correlated with relapse-free survival in patients only treated with tamoxifen. Multivariate analysis revealed that the GPER expression was an independent prognosticator for a poor outcome. In a study of postmenopausal lymph node negative breast cancer patients, the absence of the plasma membrane GPER predicted a 91% 20-year distant disease-free survival, as compared to 73% in the presence of GPER for the tamoxifen-treated ER-positive and PR-positive subgroup [20]. The GPER overexpression and plasma membrane localization are critical events in breast cancer progression. GPER was also prevalent in the TNBC, and the GPER expression was associated with a younger age and a more aggressive disease [37].

As compared with the corresponding primary tumors in the same patients, GPER expression in the recurrent tumors or metastases significantly increased under the tamoxifen treatment [27,28].

The tissue microarrays from the formalin-fixed, paraffin embedded samples of the primary invasive breast carcinomas suggested that the predominantly cytoplasmic or nuclear GPER expression were two distinct immunohistochemical patterns and may reflect different biological features [19]. Cytoplasmic GPER expression was associated with non-ductal histology, lower stage, more differentiation, and better overall survival, whereas the nuclear GPER expression was associated with poor differentiation and TNBC. In the breast cancer cell lines, confocal microscopy revealed the different GPER expression patterns. The T47D cells had a strong GPER expression, predominantly localized in the cytoplasm. The MCF-7 cells showed a less strong GPER expression and a mainly nuclear distribution. No distinct plasma membranous expression was observed.

Briefly, GPER was prevalent in the ERα positive breast cancer and TNBC. The GPER was a prognosticator for a poor outcome. There was a higher GPER expression in the re-biopsy specimen of tamoxifen resistant ERα positive breast cancer and chemotherapy refractory TNBC than the primary tumor. It should be noted that the subcellular location of the GPER may have a different prognostic implication in breast cancer.

3. G-Protein Coupled Estrogen Receptor Functions in Breast Cancer

The 17β-estradiol activated the extracellular regulated protein kinase 1/2 (ERK1/2), not only in the ERα-positive and the ERβ-positive MCF-7 cells, but also in the ERα-negative and the ERβ-negative SkBr3 cells [47]. Immunoblot analysis showed that this estrogen response was associated with the presence of the GPER protein in these cells. The ERα-negative, ERβ-positive MDA-MB-231 cells are GPER deficient and insensitive to ERK1/2 activation by E_2. Transfection of the MDA-MB-231 cells with a GPER complementary DNA resulted in the overexpression of a GPER protein and conversion to an estrogen-responsive phenotype. In addition, the GPER-dependent ERK1/2 activation could be triggered by the ER antagonist, fulvestrant. The E_2 signaling to the ERK1/2 occurred via a Gβγ-dependent, pertussis toxin-sensitive pathway. The β and γ subunits of the G protein activate the steroid receptor coactivator (SRC) tyrosine kinase, which binds to the integrin αvβ1 through the SHC adapter protein (Figure 3). The complex activates the matrix metalloproteinase, which then cleaves the pro-heparin-binding EGF-like growth factor and releases the heparin-binding EGF-like growth factor (HB-EGF) into the extracellular space. The free HB-EGF then transactivates the EGFR. The E_2 signaling to the ERK1/2 could be blocked by down-modulating the HB-EGF from the cell surface with the diphtheria toxin mutant, CRM-197, neutralizing the HB-EGF with antibodies, or inhibiting the EGFR tyrosine kinase activity. ER-negative breast cancers that continue to express GPER may use estrogen to drive the EGFR-dependent cellular responses [47,50,51]. The crosstalk between the GPER and the EGFR was confirmed in the tamoxifen-resistant ERα positive cell, TNBC, and cancer-associated fibroblast (CAF), respectively (Table 1) [26,28,38,66,67].

Table 1. G-protein coupled estrogen receptor (GPER) as a prognosticator in breast cancer cell lines and tissues.

References	Materials	Methods	Subcellular Localizations	Effects on Tumor
Tamoxifen-resistant ERα positive cells				
Ignatov 2010 [26]	MCF-7, TAM-R MCF-7	Western blot	membrane/endoplasmic reticulum	promoting
Ignatov 2011 [27]	TAM-R cancer tissue	immunohistochemistry	nucleus/cytoplasm	promoting
Mo 2013 [28]	MCF-7, TAM-R MCF-7 TAM-R cancer tissue TAM-R mouse xenograft	immunohistochemistry immunofluorescence RT-PCR, Western blot	membrane/cytoplasm	promoting
Chen 2014 [29]	MCF-7, SkBr3 cells	qRT-PCR, Western blot	non-specified	promoting
Catalano 2014 [30]	MCF-7, TAM-R MCF-7, SkBr3, CAF	RT-PCR, Western blot for aromatase activity	non-specified	promoting
Triple-negative breast cancer cells				
Lappano 2010 [35]	MCF-7, SkBr3	RT-PCR, Western blot	non-specified	promoting
Girgert 2012 [36]	MDA-MB-435, HCC1806	RT-PCR, Western blot	non-specified	promoting
Steiman 2013 [37]	TNBC cancer tissue	immunohistochemistry	non-specified	promoting
Yu 2014 [38]	MDA-MB-468, MDA-MB-436 TNBC cancer tissue	immunohistochemistry immunofluorescence RT-PCR, Western blot	nucleus/cytoplasm	promoting
Zhou 2016 [39]	SkBr3, MDA-MB-231	Western blot immunofluorescence	nucleus/cytoplasm	promoting
Albanito 2008 [40]	SkBr3, BT20	RT-PCR, Western blot immunofluorescence	nucleus/cytoplasm	promoting
Cancer-associated fibroblast				
Luo 2014, 2016 [66,67]	CAFs isolated from surgical specimens	RT-PCR, Western blot immunofluorescence	nucleus/cytoplasm	promoting
Conflicting results				
Broselid 2013 [55]	ER-(+) cancer tissue MCF-7 ± GPER knockdown T47D, HEK ± GPER	RT-PCR, Western blot immunofluorescence	non-specified	suppressive
Poola 2008 [56]	ER-(+)&(−) cancer tissue	qRT-PCR	non-specified	suppressive
Kuo 2007 [57]	ER-(+)&(−) cancer tissue	qPCR	non-specified	suppressive
Filardo 2002 [58]	MCF-7, SkBr3, MDA-MB-231	Western blot	non-specified	suppressive
Ariazi 2010 [59]	ER-(+)/(−) cancer microarray MCF-7, SkBr3	RT-PCR, Western blot Ca^{2+} imaging	non-specified	suppressive
Weißenborn 2014 [60]	MCF-7, SkBr3	RT-PCR, Western blot methylation PCR bioinformatic	non-specified	suppressive
Weißenborn 2014 [61]	MDA-MB-231, MDA-MB-468	RT-PCR, Western blot methylation PCR bioinformatic	non-specified	suppressive
Chen 2016 [62]	MDA-MB-231 TNBC cancer tissue MDA-MB-231 mice xenograft	qRT-PCR, Western blot immunofluorescence	nucleus/cytoplasm	suppressive
Liang 2017 [63]	MDA-MB-231 TNBC tissue microarray MDA-MB-231 mice xenograft	qRT-PCR, Western blot immunofluorescence	nucleus/cytoplasm	suppressive
Okamoto 2016 [64]	SkBr3 cells	qRT-PCR, Western blot	non-specified	suppressive

ER, estrogen receptor; TAM-R, tamoxifen-resistant; TNBC, triple-negative breast cancer; CAF, cancer-associated fibroblast; qRT-PCR, quantitative RT-PCR.

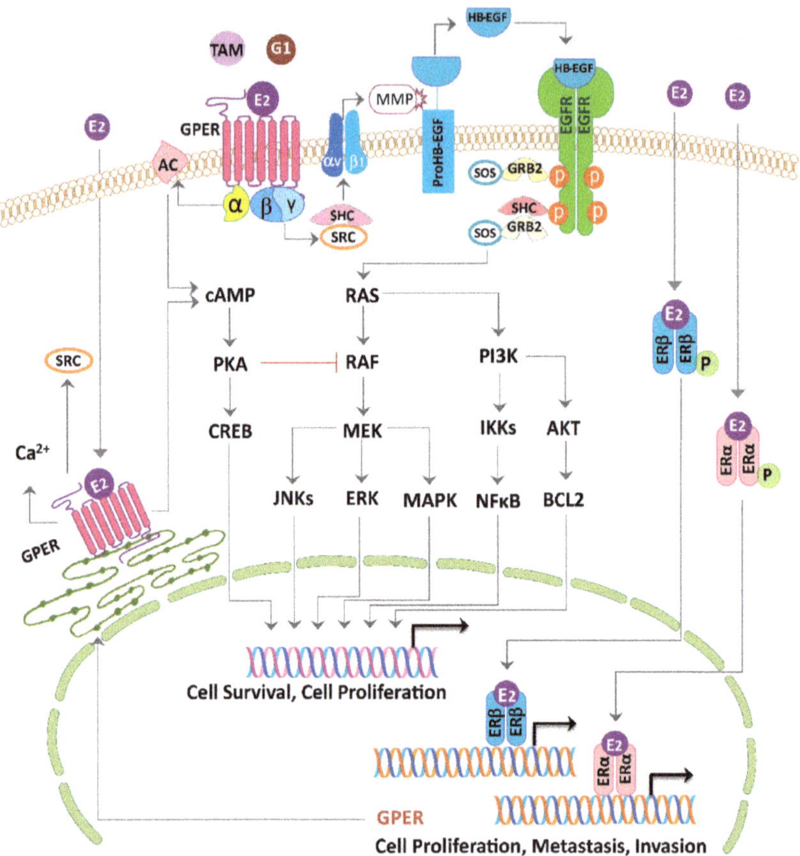

Figure 3. Model of estrogen signaling pathways in cancer. 17β-estradiol (E_2) activates ERα or ERβ to induce the receptor dimerization, and subsequently acts as a transcription factor or interacts with other transcription factors binding to the promoter region of the target genes. E_2, tamoxifen (TAM) or G1 activate the G-protein Coupled Estrogen Receptor (GPER) distributed in the nucleus, cytoplasm, and plasma membrane. Activation of GPER located in the plasma membrane stimulates steroid receptor coactivator (SRC) through a Gβγ-subunit protein pathway. The β and γ subunits of the G protein activate the SRC tyrosine kinase, which binds to the integrin αvβ1 through the SHC adapter protein. The complex activates the matrix metalloproteinase (MMP), which then cleaves the pro-heparin-binding EGF-like growth factor (proHB-EGF) and releases the heparin-binding EGF-like growth factor (HB-EGF) into the extracellular space. The free HB-EGF then transactivates the epidermal growth factor receptor (EGFR). Phosphorylation of EGFR in turn activates the downstream pathways, which can induce rapid non-genomic effects, or genomic effects regulating different genes transcription and leads to cell survival and proliferation. On the other hand, through GPER, E_2, tamoxifen or G1 is able to stimulate the adenylyl cyclase activity through a Gα-subunit protein pathway, which then leads to the protein kinase A (PKA)-mediated suppression of the EGFR-induced ERK activity. Thus, via the GPER, E_2, tamoxifen or G1 may balance the ERK activity by stimulating two distinct G-protein signaling pathways that have opposing effects on the EGFR-to-MAPK axis. Long-term tamoxifen treatment could sensitize the cancer cells through E_2-stimulated upregulation of GPER and translocation from the endoplasmic reticulum to the plasma membrane.

3.1. Tamoxifen-Resistant ERα Positive Cells

The tamoxifen-resistant cells, TAM-R, exhibited an enhanced sensitivity to the E_2 and the GPER-specific agonist, G1 (Figure 1), when compared to the parental MCF-7 cells [26,28]. Tamoxifen was able to stimulate the mitogen-activated protein kinase (MAPK) phosphorylation and cell growth in the TAM-R cells, and the effects were abolished by the GPER antagonist, G15 (Figure 1), the GPER anti-sense oligonucleotide, the selective SRC inhibitor PP2, and the EGFR inhibitor AG1478. The basal EGFR expression was only slightly elevated in the TAM-R cells, and the basal GPER expression, phosphorylation of the Ak strain transforming murine thymoma viral oncogene (AKT), and the MAPK remained unchanged when compared to the parental cells. Continuous treatment of the MCF-7 cells with G1 mimics the long-term treatment with tamoxifen and drastically increases its agonistic activity. Interestingly, the estrogen treatment significantly increased the GPER plasma membrane translocation, which was stronger in the TAM-R cells. The GPER plasma membrane translocation facilitated the crosstalk with the EGFR. The results have suggested the importance of the GPER and EGFR transactivation in the development of tamoxifen resistance. Combined therapy with the G15 and tamoxifen promoted apoptosis in a TAM-R xenograft and inhibited the drug-resistant tumor progression. The GPER activation led to the nuclear translocation of the forkhead box O3a, FOXO3a, and the down-regulation of caspase 3 and caspase 7 via the phosphoinositide 3-kinase (PI3K)/AKT pathway in the MCF-7 cells [52].

3.2. Triple-Negative Breast Cancer Cells

The GPER was strongly expressed in the TNBC cell lines, MDA-MB-468, and MDA-MB-436 [38]. Treatment with E_2, tamoxifen, and the GPER-specific agonist, G1 led to the rapid activation of ERK1/2, but not AKT. The estrogen/GPER/ERK signaling pathway was involved in the increased cell growth, survival, migration, and invasion through upregulating the expression of cyclin A, cyclin D1, Bcl-2, and c-fos that were associated with the cell cycle, anti-apoptosis, and proliferation, respectively. Pretreatment with the GPER antagonist, G15, AG1478, the ERK1/2 inhibitor, U0126, or the transfection with the siRNA against the GPER could abolish the effects. Immunohistochemical analysis of the TNBC specimens showed a significantly stronger staining of the p-ERK1/2 in the GPER-positive tissues than the GPER-negative tissues [38]. The positivity of the GPER and p-ERK1/2 displayed a strong association with the large tumor size and advanced stage, indicating that the GPER/ERK signaling might also contribute to the tumor progression in the TNBC patients, which correlated with the in vitro experimental results. 17β-estradiol and 4-hydroxytamoxifen also increased the proliferation of another two TNBC cell lines, MDA-MB-435 and HCC1806, which was completely prevented by being transfected with siRNA against the GPER [36]. The increased activity of the SRC kinase, EGFR transactivation, and c-fos expression, was also abolished after the knock-down of the GPER expression.

3.3. Cancer-Associated Fibroblasts

The GPER was expressed in the stromal fibroblasts of the primary breast cancer tissues, and the CAFs isolated [66,67]. Tamoxifen, in addition to E_2 and the G1 activated GPER, resulted in the transient increases in the cell index, intracellular calcium, ERK1/2 phosphorylation, and promoted the CAF cell cycle progression, proliferation, and migration. These effects were blocked by the G15, AG1478, and U0126. Importantly, tamoxifen, as well as G1, increased the E_2 production in the breast CAFs via the GPER/EGFR/ERK signaling pathway when the substrate of E_2, testosterone, was added to the medium. The GPER-mediated CAF-dependent estrogenic effects in the tumor-associated stroma are more likely to contribute to breast cancer progression, especially in the tamoxifen resistance, via a positive feedback loop involving the GPER/EGFR/ERK signaling pathway and E_2 production.

17β-estradiol and G1 triggered the GPER/EGFR/ERK/c-fos signaling pathway that led to an increased vascular endothelial growth factor (VEGF) via the upregulation of the hypoxia-inducible factor-1α (HIF1α) in the ER-negative breast cancer cells and the CAFs [53]. The conditioned medium from the CAFs treated with E$_2$ and G1 promoted the human endothelial tube formation in a GPER-dependent manner. In the mice breast cancer xenograft model, GPER activation enhanced the tumor growth and the expression of HIF1α, VEGF, and the endothelial marker, CD34. Fatty acid synthase catalyzes the de novo biogenesis of the fatty acids and acts as a metabolic oncogene. 17β-estradiol and G1 regulated the fatty acid synthase expression and activity through the GPER/EGFR/ERK/c-fos/activator protein 1 (AP-1) signaling pathway in the SkBr3 cells and CAFs [54].

In summary, E$_2$, tamoxifen, and G1 upregulate estrogen production, increase GPER expression and plasma membrane translocation, and stimulate the proliferation, migration, invasion of breast cancer cell lines, and cancer-associated fibroblasts. Pretreatment with the GPER antagonist, G15 or transfected with siRNA against GPER attenuates the effects. The tumor promoting effects of the GPER operate through the EGFR transactivation and related signaling pathways.

3.4. Controversies about G-Protein Coupled Estrogen Receptor Function in Breast Cancer

Contrary to the majority of studies that have reported a tumor promoting effect of GPER activation in breast cancer, several studies have demonstrated that GPER functions as a tumor suppressor and induces apoptosis (Table 1) [55–65].

The GPER expression by immunohistochemistry had been reported as a prognosticator for the increased distant disease-free survival in patients with ER-positive breast cancer treated with tamoxifen [55]. A constitutive GPER-dependent pro-apoptotic signaling was proposed. The GPER expression at mRNA levels was significantly down-regulated in both the ERα-positive and ERα-negative breast cancer tissues in comparison with their matched normal tissues, and significantly lower in tumor tissues from the patients who had lymph node metastasis than those without [56]. The tumor samples from 118 Taiwanese patients with infiltrating ductal carcinoma of the breast had a lower GPER expression at the mRNA level than that in non-tumor mammary tissues [57]. The correlation of the GPER expression with clinical parameters and patient survival was not significant.

Filardo et al., suggested that via the GPER, estrogens as well as antiestrogens, were capable of stimulating the adenylyl cyclase activity and increasing the cAMP concentration, which in turn, led to the PKA-mediated suppression of the EGFR induced ERK1/2 activity (Figure 3) [58]. Thus, via the GPER, estrogen may balance ERK1/2 activity by stimulating two distinct G-protein signaling pathways that have opposing effects on the EGFR-to-MAPK axis. The other study concurred with the observation that the reduced cAMP generation attenuated the inhibition of EGFR signaling [28].

The GPER functions promoted the SkBr-3 but inhibited the MCF-7 cellular proliferation. An ER- and [Ca^{2+}]-dependent negative feedback was proposed for the difference [59]. 17β-estradiol is known to downregulate the ERα expression in the MCF-7 cells as a negative feedback regulatory loop to prevent overexpression. Likewise, the GPER may also be negatively regulated by E$_2$ via the ER to prevent an excessive GPER-dependent activity, such as being aberrantly high [Ca^{2+}]. The maximum increases in [Ca^{2+}] were much larger in the SkBr-3 cells than in the MCF-7 cells. It is possible that this was due to the lack of ERs in the SkBr-3 cells, which translated into a lack of negative feedback regulation.

Proliferative results were observed with two non-specific GPER agonists, estrogen and tamoxifen, but not with its specific agonist, G1. The G1-induced inhibitory effect was specific for the GPER expressing cells [60,61]. Radiation induced different changes in the GPER expression among the breast cancer cells; upregulated in the MDA-MB-231 and the MDA-MB-468 cells and down-regulated in the MCF-7 cells. They proposed that it was linked to the form of the p53 protein expressed in each cell. While the MDA-MB-231 and the MDA-MB-468 cells express a non-functional p53 gene, the MCF-7 cells have normal wild-type p53. A feedback mechanism exists between the expression of p53 and

GPER. The GPER activation increases the p53 expression, which in turn down-regulates the expression of the GPER [61].

The activation of the GPER suppresses the epithelial mesenchymal transition, migration, and angiogenesis of the TNBC via the nuclear factor-kappa B (NF-κB) signals [62,63]. The GPER also inhibits the tumor necrosis factor alpha-induced expression of interleukin 6 through the repression of the NF-κB promoter activity in the SkBr3 cells [64].

Further studies are therefore necessary to define the role of the GPER, proliferative or pro-apoptotic in breast cancer.

4. G-protein Coupled Estrogen Receptor Knockout Mice

In contrast with the pharmacological methods, the GPER knockout more clearly understands the GPER function through targeted gene deletion or disruption [17]. Four GPER knockout mice have been reported [68–71]. Among them, three have the whole GPER coding region deleted, and the fourth with the C-terminal portion of the GPER remaining. Although the GPER was also distributed in normal breast tissues, histopathological analysis did not reveal any abnormalities in the GPER-knockout mice [68]. Mammary gland responses after estradiol with/without progesterone treatment were also unimpaired in the GPER knockout mice. The GPER knockout mice did not show overt phenotypes in viability or reproductive function, but some functions of estrogen have been absent in the GPER knockout mice that support the GPER as a physiologically relevant estrogen receptor [17,68–72]. In the only study of breast cancer, the GPER knockout in the polyoma middle T antigen-mouse mammary tumor virus transgenic mice revealed smaller tumors and reduced metastasis [18].

5. Interactions between Cancer Cells and Cancer-Associated Fibroblasts through G-Protein Coupled Estrogen Receptor

An increased aromatase expression and activity was found in the tamoxifen resistant breast cancer cells [30]. Knocking-down the GPER expression reversed the enhanced aromatase levels. In the ER-negative, GPER/aromatase-positive SkBr3 cells, tamoxifen acted as a GPER agonist. The tamoxifen treatment increased the aromatase expression through an enhanced recruitment of the c-fos/c-jun complex to the AP-1 responsive elements located within the promoter region. Tamoxifen also induced aromatase expression via the GPER in the CAFs. The increased estrogen production in the microenvironment may well lead to a more aggressive behavior of breast cancers.

On co-culturing the CAFs with the breast cancer cells, a significant GPER translocation from the nucleus to the cytoplasm was observed in the CAFs, similar to that observed in the stromal fibroblast in the breast cancer tissues, indicating that the cancer cells may affect the subcellular localization of the GPER in the CAFs. CRM1, a nuclear export protein, and activated PI3K/AKT signaling pathway are involved in the cytoplasmic GPER translocation in the CAFs, which in turn activates a novel estrogen/GPER/cyclic AMP (cAMP)/protein kinase A (PKA)/cAMP response element binding protein signaling pathway, and triggers aerobic glycolysis in the CAFs [73]. The glycolytic CAFs feed pyruvate and lactate to cancer cells to undergo oxidative phosphorylation and contribute to drug resistance. The stromal GPER-mediated drug resistance from the reprogramming of the tumor energy metabolism, i.e. the "reverse Warburg effect", provided the rationale for the CAFs as a promising target for therapy [74,75]. The different subcellular location of the GPER in the breast CAFs may have biological implications [76]. Targeting the cytoplasmic GPER in the CAFs may restore the response to treatments in both the ER-positive and -negative breast cancers.

6. Future Perspectives in the Study of the G-Protein Coupled Estrogen Receptor in Breast Cancer

The inconsistent observations among these studies could be attributed to the usage of the different subtypes of breast cancer samples and the different subcellular localization of the GPER [60], the difference of the cell types and treatment conditions, and the specificities of the agonist [61]. The specificity of the GPER antibodies used in the immunohistochemistry and Western blot may affect the staining patterns of the GPER [77]. The epigenetic of GPER, tumor microenvironment, and hormone levels also affected the results [65].

Systemic approaches via epidemiology and laboratory studies are necessary to confirm the role of GPER in breast cancer in the future. ERα positive breast cancer and TNBC need to be studied separately. What is the true prevalence of GPER in the ERα positive breast cancer and the TNBC, respectively? The GPER expression of the archival tissue in both tumor and associated stroma should be measured by a unified scoring system including the staining intensity and the proportion of positive cancer cells, as per the Allred scoring for ER [78]. The subcellular location of the GPER, i.e. nucleus, cytoplasm, and plasma membrane, also need be notified to clarify the implication of the translocation and its role in the prognosis.

The GPER expression needs to be compared between the tumors and their matched normal tissues. The association of GPER with the ERα, PR, HER2/neu, and the correlations with the clinic-pathologic variables and survival also need to be investigated. In addition to stage, histology, and ER, PR, HER2/neu status, the smoking history, and menopausal status should be included in a multivariate survival analysis to determine as to whether the GPER is an independent prognosticator. Does high or low GPER expression before treatment predict the development of tamoxifen resistance or refractory TNBC? Pairwise comparison of the GPER expression by immunohistochemistry between the archival tissues of the treatment-naïve and re-biopsy specimen of tamoxifen resistant ERα positive breast cancer or the chemotherapy refractory TNBC, will help to better understand whether the GPER becomes predominant during treatment.

The baseline GPER expression in the ERα positive, tamoxifen-resistant ERα positive, TNBC cell lines, and the CAFs, their differences including the subcellular location, the estrogen production including the aromatase activity, the subsequent GPER expression, and the effects, i.e. proliferation, migration, invasion after treatment with E_2, tamoxifen and G1, and whether the effects are abolished by pretreatment with G15 or siRNA against the GPER will help to confirm the role of GPER. The differences between the effects of E_2, tamoxifen and G1 should also be observed for the receptor specificity. Immunofluorescent microscopy may be used to observe the cell surface translocation of the GPER, which facilitates the crosstalk with EGFR. In addition to the ERK signaling pathway, is the AKT pathway involved in the GPER-EGFR transactivation [38]? The reciprocal changes of the GPER expression, the biomarkers for the "reverse Warburg effect", and the lactate shuttle such as the mono-carboxylate transporters 4, and the mitochondrial activities in the cancer cell lines and the CAFs, on treatment with E_2, tamoxifen, or G1 also deserve to be studied, to explore the role of the metabolic coupling that occurs between the CAFs and the cancer cells.

For the reported anti-proliferative and pro-apoptotic effects of the GPER activation in the literature, the stimulation of the adenylyl cyclase to increase the intracellular cAMP or intracellular Ca^{2+} mobilization as a second messenger, and the pro-apoptotic signaling, measured by the increased cytochrome C release, caspase-3 cleavage, poly(ADP-ribose) polymerase, PARP cleavage, and the decreased cell viability after treatment with estrogen, tamoxifen, G1, or G15, could be correlated to understand the regulatory mechanism [58]. In contrast with the chronic exposure at a low nanomolar dosage, the higher bisphenol concentration in the micromolar exerts an anti-proliferative effect on the cancer cells in spite of the activation of the EGFR/ERK signaling pathway. An increased expression of p53 and its phosphorylation was observed. Therefore, we need to investigate whether the GPER effects are modified by the simultaneous p53 activation [79].

Several other signaling pathways, such as HIPPO, NOTCH, and target genes and proteins, such as SNAIL, β1-integrin, focal adhesion kinase, calpain, and connective tissue growth factor, have been reported to be involved in the GPER-mediated breast cancer progression [29,80–83]. The microRNAs target numerous genes and are involved in cancer progression. GPER is an important regulator of microRNAs. In the MCF-7 and MDA-MB-231 cells, the GPER activation down-regulated the miR-148a and caused an increase in the human leukocyte antigen-C expression that led to the cancer cells escaping from immune surveillance and allowed cancer progression [84]. In the SkBr-3 cells and

CAFs, the GPER activation up-regulated the miR-144, which in turn inhibited the tumor suppressor runt-related transcription 1 factor, and increased the cell cycle propagation [85]. Systemic studies of microRNA expression are needed to better define the regulations by the GPER and their effects.

To include GPER in the functional screens for genes contributing to tamoxifen resistance in breast cancer cells and the usage of novel technology, e.g., DNA microarray, and proteomic analysis may help to investigate the association [86,87]. Insulin-like growth factor-I and HIF1α were shown to increase the GPER expression in the breast cancer cells and led to cell proliferation, migration, and tumor angiogenesis [88,89]. The epidermal growth factor may reciprocally up-regulate the GPER to facilitate a stimulatory role of estrogen, even in the TNBC [40]. How the GPER expression is regulated, e.g., through epigenetic by methylation or demethylation of the promoter, requires further studies to understand the EGFR and GPER crosstalk [61,90].

The classical ERs do not contain a hydrophobic part that may serve as a transmembrane domain. However, the presence of the ERs in the membrane of the somatic and cancer cells, and the rapid non-genomic responses that occur through the membrane-bound classical ERα and ERβ have been reported [49,91–93]. How the classical ERs translocate to the membrane [94], and the interactions between the classical ERs and the GPER is important to understand the membrane-associated non-genomic pathways of estrogen.

7. Conclusions

In this review, we have tried to explore the actions and to understand the molecular basis of the agonist/antagonist mechanisms of the GPER in breast cancer with tamoxifen resistance, and TNBC from the current epidemiology and laboratory studies. The combined non-genomic and genomic effects of estrogen are critical for its overall function, even in the absence of ligand, and the interactions between multiple receptors are complex [95]. Further studies will help to clarify the role of the GPER and support it as a novel target of therapeutic strategies. GPER expression may also be valuable as prognostic or predictive biomarkers.

Author Contributions: L.-H.H. and S.-H.K. conceived the paper; L.-H.H. and S.-H.K. wrote the paper; L.-H.H., N.-M.C., Y.-F.L., and S.-H.K. revised the paper.

Funding: This research received no external funding.

Acknowledgments: The authors would like to offer their sincere thanks to Michael Wise for help with English language editing; cytotechnologists Jen-Sheng Ko and Mei-Ling Wu for the figures preparation; Shiao-Chiu Huang for the assistance with the references preparation.

Conflicts of Interest: The authors declare no conflicts of interest.

Abbreviations

AKT	Ak strain transforming murine thymoma viral oncogene
AP-1	activator protein 1
CAF	cancer-associated fibroblast
cAMP	cyclic AMP
E_2	17β-estradiol
EGFR	epidermal growth factor receptor
ER	estrogen receptor
ERα	estrogen receptor α
ERβ	estrogen receptor β
ERK	extracellular regulated protein kinase
GPER	G-protein coupled estrogen receptor
HB-EGF	heparin-binding EGF-like growth factor
HER2/neu	human epidermal growth factor receptor 2
HIF1α	hypoxia-inducible transcription factor-1α
MAPK	mitogen-activated protein kinase

NF-κB	nuclear factor-kappa B
PI3K	Phosphoinositide 3-kinase
PKA	protein kinase A
PR	progesterone receptor
SERM	selective estrogen receptor modulator
SRC	steroid receptor coactivator
TNBC	triple-negative breast cancer
VEGF	vascular endothelial growth factor

References

1. Kamangar, F.; Dores, G.M.; Anderson, W.F. Patterns of cancer incidence, mortality, and prevalence across five continents: Defining priorities to reduce cancer disparities in different geographic regions of the world. *J. Clin. Oncol.* **2006**, *24*, 2137–2150. [CrossRef] [PubMed]
2. Siegel, R.L.; Miller, K.D.; Jemal, A. Cancer statistics, 2018. *CA Cancer J. Clin.* **2018**, *68*, 7–30. [CrossRef] [PubMed]
3. Health Promotion Administration, Ministry of Health and Welfare, Taiwan Cancer Registry Annual Report, 2015, Taiwan. Available online: http://www.hpa.gov.tw (accessed on 21 November 2018).
4. NCCN Clinical Practice Guidelines in Oncology: Breast Cancer Risk Reduction. Available online: http://www.nccn.org (accessed on 21 November 2018).
5. Carmeci, C.; Thompson, D.A.; Ring, H.Z.; Francke, U.; Weigel, R.J. Identification of a gene (GPR30) with homology to the G-protein-coupled receptor superfamily associated with estrogen receptor expression in breast cancer. *Genomics* **1997**, *45*, 607–617. [CrossRef] [PubMed]
6. Revankar, C.M.; Cimino, D.F.; Sklar, L.A.; Arterburn, J.B.; Prossnitz, E.R. A transmembrane intracellular estrogen receptor mediates rapid cell signaling. *Science* **2005**, *307*, 1625–1630. [CrossRef] [PubMed]
7. Thomas, P.; Pang, Y.; Filardo, E.J.; Dong, J. Identity of an estrogen membrane receptor coupled to a G protein in human breast cancer cells. *Endocrinology* **2005**, *146*, 624–632. [CrossRef] [PubMed]
8. Filardo, E.; Quinn, J.; Pang, Y.; Graeber, C.; Shaw, S.; Dong, J.; Thomas, P. Activation of the novel estrogen receptor G protein-coupled receptor 30 (GPR30) at the plasma membrane. *Endocrinology* **2007**, *148*, 3236–3245. [CrossRef]
9. Dennis, M.K.; Burai, R.; Ramesh, C.; Petrie, W.K.; Alcon, S.N.; Nayak, T.K.; Bologa, C.G.; Leitao, A.; Brailoiu, E.; Deliu, E.; et al. In vivo effects of a GPR30 antagonist. *Nat. Chem. Biol.* **2009**, *5*, 421–427. [CrossRef]
10. Pupo, M.; Maggiolini, M.; Musti, A.M. GPER mediates non-genomic effects of estrogen. *Methods Mol. Biol.* **2016**, *1366*, 471–488.
11. Rae, J.M.; Johnson, M.D. What does an orphan G-protein-coupled receptor have to do with estrogen? *Breast Cancer Res.* **2005**, *7*, 243–244. [CrossRef]
12. Lappano, R.; Rosano, C.; Santolla, M.F.; Pupo, M.; De Francesco, E.M.; De Marco, P.; Ponassi, M.; Spallarossa, A.; Ranise, A.; Maggiolini, M. Two novel GPER agonists induce gene expression changes and growth effects in cancer cells. *Curr. Cancer Drug Targets* **2012**, *12*, 531–542. [CrossRef]
13. Méndez-Luna, D.; Bello, M.; Correa-Basurto, J. Understanding the molecular basis of agonist/antagonist mechanism of GPER1/GPR30 through structural and energetic analyses. *Steroid Biochem. Mol. Biol.* **2016**, *158*, 104–116. [CrossRef] [PubMed]
14. Aiello, F.; Carullo, G.; Giordano, F.; Spina, E.; Nigro, A.; Garofalo, A.; Tassini, S.; Costantino, G.; Vincetti, P.; Bruno, A.; et al. Identification of breast cancer inhibitors specific for G protein-coupled estrogen receptor (GPER)-expressing cells. *ChemMedChem* **2017**, *12*, 1279–1285. [CrossRef] [PubMed]
15. Filardo, E.J.; Graeber, C.T.; Quinn, J.A.; Resnick, M.B.; Giri, D.; DeLellis, R.A.; Steinhoff, M.M.; Sabo, E. Distribution of GPR30, a seven membrane-spanning estrogen receptor, in primary breast cancer and its association with clinicopathologic determinants of tumor progression. *Clin. Cancer Res.* **2006**, *12*, 6359–6366. [CrossRef] [PubMed]

16. Marot, D.; Bieche, I.; Aumas, C.; Esselin, S.; Bouquet, C.; Vacher, S.; Lazennec, G.; Perricaudet, M.; Kuttenn, F.; Lidereau, R.; et al. High tumoral levels of Kiss1 and G-protein-coupled receptor 54 expression are correlated with poor prognosis of estrogen receptor-positive breast tumors. *Endocr. Relat. Cancer* **2007**, *14*, 691–702. [CrossRef] [PubMed]

17. Prossnitz, E.R.; Hathaway, H.J. What have we learned about GPER function in physiology and disease from knockout mice? *J. Steroid Biochem. Mol. Biol.* **2015**, *153*, 114–126. [CrossRef]

18. Marjon, N.A.; Hu, C.; Hathaway, H.J.; Prossnitz, E.R. G protein-coupled estrogen receptor regulates mammary tumorigenesis and metastasis. *Mol. Cancer Res.* **2014**, *12*, 1644–1654. [CrossRef]

19. Samartzis, E.P.; Noske, A.; Meisel, A.; Varga, Z.; Fink, D.; Imesch, P. The G protein-coupled estrogen receptor (GPER) is expressed in two different subcellular localizations reflecting distinct tumor properties in breast cancer. *PLoS ONE* **2014**, *9*, e83296. [CrossRef] [PubMed]

20. Sjöström, M.; Hartman, L.; Grabau, D.; Fornander, T.; Malmström, P.; Nordenskjöld, B.; Sgroi, D.C.; Skoog, L.; Stål, O.; Leeb-Lundberg, L.M.; et al. Lack of G protein-coupled estrogen receptor (GPER) in the plasma membrane is associated with excellent long-term prognosis in breast cancer. *Breast Cancer Res. Treat.* **2014**, *145*, 61–71. [CrossRef] [PubMed]

21. Filardo, E.J. A role for G-protein coupled estrogen receptor (GPER) in estrogen-induced carcinogenesis: Dysregulated glandular homeostasis, survival and metastasis. *J. Steroid Biochem. Mol. Biol.* **2018**, *176*, 38–48. [CrossRef]

22. Wang, D.; Hu, L.; Zhang, G.; Zhang, L.; Chen, C. G protein-coupled receptor 30 in tumor development. *Endocrine* **2010**, *38*, 29–37. [CrossRef] [PubMed]

23. Lappano, R.; Pisano, A.; Maggiolini, M. GPER function in breast cancer: An overview. *Front. Endocrinol.* **2014**, *5*, 66. [CrossRef] [PubMed]

24. Jacenik, D.; Cygankiewicz, A.I.; Krajewska, W.M. The G protein-coupled estrogen receptor as a modulator of neoplastic transformation. *Mol. Cell Endocrinol.* **2016**, *429*, 10–18. [CrossRef] [PubMed]

25. Molina, L.; Figueroa, C.D.; Bhoola, K.D.; Ehrenfeld, P. GPER-1/GPR30 a novel estrogen receptor sited in the cell membrane: Therapeutic coupling to breast cancer. *Expert Opin. Ther. Targets* **2017**, *21*, 755–766. [CrossRef] [PubMed]

26. Ignatov, A.; Ignatov, T.; Roessner, A.; Costa, S.D.; Kalinski, T. Role of GPR30 in the mechanisms of tamoxifen resistance in breast cancer MCF-7 cells. *Breast Cancer Res. Treat.* **2010**, *123*, 87–96. [CrossRef] [PubMed]

27. Ignatov, A.; Ignatov, T.; Weissenborn, C.; Eggemann, H.; Bischoff, J.; Semczuk, A.; Roessner, A.; Costa, S.D.; Kalinski, T. G-protein-coupled estrogen receptor GPR30 and tamoxifen resistance in breast cancer. *Breast Cancer Res. Treat.* **2011**, *128*, 457–466. [CrossRef] [PubMed]

28. Mo, Z.; Liu, M.; Yang, F.; Luo, H.; Li, Z.; Tu, G.; Yang, G. GPR30 as an initiator of tamoxifen resistance in hormone-dependent breast cancer. *Breast Cancer Res.* **2013**, *15*, R114. [CrossRef] [PubMed]

29. Chen, Y.; Li, Z.; He, Y.; Shang, D.; Pan, J.; Wang, H.; Chen, H.; Zhu, Z.; Wan, L.; Wang, X. Estrogen and pure antiestrogen fulvestrant (ICI 182 780) augment cell-matrigel adhesion of MCF-7 breast cancer cells through a novel G protein coupled estrogen receptor (GPR30)-to-calpain signaling axis. *Toxicol. Appl. Pharmacol.* **2014**, *275*, 176–181. [CrossRef] [PubMed]

30. Catalano, S.; Giordano, C.; Panza, S.; Chemi, F.; Bonofiglio, D.; Lanzino, M.; Rizza, P.; Romeo, F.; Fuqua, S.A.; Maggiolini, M.; et al. Tamoxifen through GPER upregulates aromatase expression: A novel mechanism sustaining tamoxifen-resistant breast cancer cell growth. *Breast Cancer Res. Treat.* **2014**, *146*, 273–285. [CrossRef]

31. Nass, N.; Kalinski, T. Tamoxifen resistance: From cell culture experiments towards novel biomarkers. *Pathol. Res. Pract.* **2015**, *211*, 189–197. [CrossRef]

32. Rondón-Lagos, M.; Villegas, V.E.; Rangel, N.; Sánchez, M.C.; Zaphiropoulos, P.G. Tamoxifen resistance: Emerging molecular targets. *Int. J. Mol. Sci.* **2016**, *17*, 1357. [CrossRef] [PubMed]

33. Ignatov, T.; Eggemann, H.; Semczuk, A.; Smith, B.; Bischoff, J.; Roessner, A.; Costa, S.D.; Kalinski, T.; Ignatov, A. Role of GPR30 in endometrial pathology after tamoxifen for breast cancer. *Am. J. Obstet. Gynecol.* **2010**, *203*, 595. [CrossRef] [PubMed]

34. Hu, R.; Hilakivi-Clarke, L.; Clarke, R. Molecular mechanisms of tamoxifen-associated endometrial cancer. *Oncol. Lett.* **2015**, *9*, 1495–1501. [CrossRef] [PubMed]

35. Lappano, R.; Rosano, C.; De Marco, P.; De Francesco, E.M.; Pezzi, V.; Maggiolini, M. Estriol acts as a GPR30 antagonist in estrogen receptor-negative breast cancer cells. *Mol. Cell Endocrinol.* **2010**, *320*, 162–170. [CrossRef] [PubMed]

36. Girgert, R.; Emons, G.; Gründker, C. Inactivation of GPR30 reduces growth of triple-negative breast cancer cells: Possible application in targeted therapy. *Breast Cancer Res. Treat.* **2012**, *134*, 199–205. [CrossRef] [PubMed]

37. Steiman, J.; Peralta, E.A.; Louis, S.; Kamel, O. Biology of the estrogen receptor, GPR30, in triple negative breast cancer. *Am. J. Surg.* **2013**, *206*, 698–703. [CrossRef] [PubMed]

38. Yu, T.; Liu, M.; Luo, H.; Wu, C.; Tang, X.; Tang, S.; Hu, P.; Yan, Y.; Wang, Z.; Tu, G. GPER mediates enhanced cell viability and motility via non-genomic signaling induced by 17β-estradiol in triple-negative breast cancer (TNBC) cells. *J. Steroid Biochem. Mol. Biol.* **2014**, *143*, 392–403. [CrossRef]

39. Zhou, K.; Sun, P.; Zhang, Y.; You, X.; Li, P.; Wang, T. Estrogen stimulated migration and invasion of estrogen receptor-negative breast cancer cells involves an ezrin-dependent crosstalk between G protein-coupled receptor 30 and estrogen receptor beta signaling. *Steroids* **2016**, *111*, 113–120. [CrossRef]

40. Albanito, L.; Sisci, D.; Aquila, S.; Brunelli, E.; Vivacqua, A.; Madeo, A.; Lappano, R.; Pandey, D.P.; Picard, D.; Mauro, L.; et al. Epidermal growth factor induces G protein-coupled receptor 30 expression in estrogen receptor-negative breast cancer cells. *Endocrinology* **2008**, *149*, 3799–3808. [CrossRef]

41. Lapensee, E.W.; Tuttle, T.R.; Fox, S.R.; Ben-Jonathan, N. Bisphenol A at low nanomolar doses confers chemoresistance in estrogen receptor-alpha-positive and -negative breast cancer cells. *Environ. Health Perspect.* **2009**, *117*, 175–180. [CrossRef]

42. Dong, S.; Terasaka, S.; Kiyama, R. Bisphenol A induces a rapid activation of Erk1/2 through GPR30 in human breast cancer cells. *Environ. Pollut.* **2011**, *159*, 212–218. [CrossRef]

43. Yu, X.; Filardo, E.J.; Shaikh, Z.A. The membrane estrogen receptor GPR30 mediates cadmium-induced proliferation of breast cancer cells. *Toxicol. Appl. Pharmacol.* **2010**, *245*, 83–90. [CrossRef] [PubMed]

44. Song, H.; Zhang, T.; Yang, P.; Li, M.; Yang, Y.; Wang, Y.; Du, J.; Pan, K.; Zhang, K. Low doses of bisphenol A stimulate the proliferation of breast cancer cells via ERK1/2/ERRγ signals. *Toxicol. Vitro* **2015**, *30*, 521–528. [CrossRef] [PubMed]

45. Lei, B.; Peng, W.; Xu, G.; Wu, M.; Wen, Y.; Xu, J.; Yu, Z.; Wang, Y. Activation of G protein-coupled receptor 30 by thiodiphenol promotes proliferation of estrogen receptor α-positive breast cancer cells. *Chemosphere* **2017**, *169*, 204–211. [CrossRef] [PubMed]

46. Biscardi, J.S.; Ishizawar, R.C.; Silva, C.M.; Parsons, S.J. Tyrosine kinase signalling in breast cancer: Epidermal growth factor receptor and c-Src interactions in breast cancer. *Breast Cancer Res.* **2000**, *2*, 203–210. [CrossRef] [PubMed]

47. Filardo, E.J.; Quinn, J.A.; Bland, K.I.; Frackelton, A.R., Jr. Estrogen-induced activation of Erk-1 and Erk-2 requires the G protein-coupled receptor homolog, GPR30, and occurs via trans-activation of the epidermal growth factor receptor through release of HB-EGF. *Mol. Endocrinol.* **2000**, *14*, 1649–1660. [CrossRef] [PubMed]

48. Filardo, E.J. Epidermal growth factor receptor (EGFR) transactivation by estrogen via the G-protein-coupled receptor, GPR30: A novel signaling pathway with potential significance for breast cancer. *J. Steroid Biochem. Mol. Biol.* **2002**, *80*, 231–238. [CrossRef]

49. Levin, E.R. Bidirectional signaling between the estrogen receptor and the epidermal growth factor receptor. *Mol. Endocrinol.* **2003**, *17*, 309–317. [CrossRef]

50. Filardo, E.J.; Thomas, P. GPR30: A seven-transmembrane-spanning estrogen receptor that triggers EGF release. *Trends Endocrinol. Metab.* **2005**, *16*, 362–367. [CrossRef]

51. Quinn, J.A.; Graeber, C.T.; Frackelton, A.R., Jr.; Kim, M.; Schwarzbauer, J.E.; Filardo, E.J. Coordinate regulation of estrogen-mediated fibronectin matrix assembly and epidermal growth factor receptor transactivation by the G protein-coupled receptor, GPR30. *Mol. Endocrinol.* **2009**, *23*, 1052–1064. [CrossRef]

52. Zekas, E.; Prossnitz, E.R. Estrogen-mediated inactivation of FOXO3a by the G protein-coupled estrogen receptor GPER. *BMC Cancer* **2015**, *15*, 702. [CrossRef]

53. De Francesco, E.M.; Pellegrino, M.; Santolla, M.F.; Lappano, R.; Ricchio, E.; Abonante, S.; Maggiolini, M. GPER mediates activation of HIF1α/VEGF signaling by estrogens. *Cancer Res.* **2014**, *74*, 4053–4064. [CrossRef] [PubMed]

54. Santolla, M.F.; Lappano, R.; De Marco, P.; Pupo, M.; Vivacqua, A.; Sisci, D.; Abonante, S.; Iacopetta, D.; Cappello, A.R.; Dolce, V.; et al. G protein-coupled estrogen receptor mediates the up-regulation of fatty acid synthase induced by 17β-estradiol in cancer cells and cancer-associated fibroblasts. *J. Biol.Chem.* **2012**, *287*, 43234–43245. [CrossRef] [PubMed]

55. Broselid, S.; Cheng, B.; Sjöström, M.; Lövgren, K.; Klug-De Santiago, H.L.; Belting, M.; Jirström, K.; Malmström, P.; Olde, B.; Bendahl, P.O.; et al. G protein-coupled estrogen receptor is apoptotic and correlates with increased distant disease-free survival of estrogen receptor-positive breast cancer patients. *Clin. Cancer Res.* **2013**, *19*, 1681–1692. [CrossRef] [PubMed]

56. Poola, I.; Abraham, J.; Liu, A.; Marshalleck, J.J.; Dewitty, R.L. The cell surface estrogen receptor, G protein-coupled receptor 30 (GPR30), is markedly down regulated during breast tumorigenesis. *Breast Cancer* **2008**, *1*, 65–78. [CrossRef] [PubMed]

57. Kuo, W.H.; Chang, L.Y.; Liu, D.L.; Hwa, H.L.; Lin, J.J.; Lee, P.H.; Chen, C.N.; Lien, H.C.; Yuan, R.H.; Shun, C.T.; et al. The interactions between GPR30 and the major biomarkers in infiltrating ductal carcinoma of the breast in an Asian population. *Taiwan J. Obstet. Gynecol.* **2007**, *46*, 135–145. [CrossRef]

58. Filardo, E.J.; Quinn, J.A.; Frackelton, A.R., Jr.; Bland, K.I. Estrogen action via the G protein-coupled receptor, GPR30: Stimulation of adenylyl cyclase and cAMP-mediated attenuation of the epidermal growth factor receptor-to-MAPK signaling axis. *Mol. Endocrinol.* **2002**, *16*, 70–84. [CrossRef]

59. Ariazi, E.A.; Brailoiu, E.; Yerrum, S.; Shupp, H.A.; Slifker, M.J.; Cunliffe, H.E.; Black, M.A.; Donato, A.L.; Arterburn, J.B.; Oprea, T.I.; et al. The G protein-coupled receptor GPR30 inhibits proliferation of estrogen receptor-positive breast cancer cells. *Cancer Res.* **2010**, *70*, 1184–1194. [CrossRef]

60. Weißenborn, C.; Ignatov, T.; Poehlmann, A.; Wege, A.K.; Costa, S.D.; Zenclussen, A.C.; Ignatov, A. GPER functions as a tumor suppressor in MCF-7 and SK-BR-3 breast cancer cells. *J. Cancer Res. Clin. Oncol.* **2014**, *140*, 663–671. [CrossRef]

61. Weißenborn, C.; Ignatov, T.; Ochel, H.J.; Costa, S.D.; Zenclussen, A.C.; Ignatova, Z.; Ignatov, A. GPER functions as a tumor suppressor in triple-negative breast cancer cells. *J. Cancer Res. Clin. Oncol.* **2014**, *140*, 713–723. [CrossRef]

62. Chen, Z.J.; Wei, W.; Jiang, G.M.; Liu, H.; Wei, W.D.; Yang, X.; Wu, Y.M.; Liu, H.; Wong, C.K.; Du, J.; et al. Activation of GPER suppresses epithelial mesenchymal transition of triple negative breast cancer cells via NF-κB signals. *Mol. Oncol.* **2016**, *10*, 775–788. [CrossRef]

63. Liang, S.; Chen, Z.; Jiang, G.; Zhou, Y.; Liu, Q.; Su, Q.; Wei, W.; Du, J.; Wang, H. Activation of GPER suppresses migration and angiogenesis of triple negative breast cancer via inhibition of NF-κB/IL-6 signals. *Cancer Lett.* **2017**, *386*, 12–23. [CrossRef] [PubMed]

64. Okamoto, M.; Mizukami, Y. GPER negatively regulates TNFα-induced IL-6 production in human breast cancer cells via NF-κB pathway. *Endocr. J.* **2016**, *63*, 485–493. [CrossRef] [PubMed]

65. Qian, H.; Xuan, J.; Liu, Y.; Shi, G. Function of G-protein-coupled estrogen receptor-1 in reproductive system tumors. *J. Immunol. Res.* **2016**, *2016*, 7128702. [CrossRef] [PubMed]

66. Luo, H.; Yang, G.; Yu, T.; Luo, S.; Wu, C.; Sun, Y.; Liu, M.; Tu, G. GPER-mediated proliferation and estradiol production in breast cancer-associated fibroblasts. *Endocr. Relat. Cancer* **2014**, *21*, 355–369. [CrossRef] [PubMed]

67. Luo, H.; Liu, M.; Luo, S.; Yu, T.; Wu, C.; Yang, G.; Tu, G. Dynamic monitoring of GPER-mediated estrogenic effects in breast cancer associated fibroblasts: An alternative role of estrogen in mammary carcinoma development. *Steroids* **2016**, *112*, 1–11. [CrossRef] [PubMed]

68. Otto, C.; Fuchs, I.; Kauselmann, G.; Kern, H.; Zevnik, B.; Andreasen, P.; Schwarz, G.; Altmann, H.; Klewer, M.; Schoor, M.; et al. GPR30 does not mediate estrogenic responses in reproductive organs in mice. *Biol. Reprod.* **2009**, *80*, 34–41. [CrossRef] [PubMed]

69. Martensson, U.E.; Salehi, S.A.; Windahl, S.; Gomez, M.F.; Sward, K.; Daszkiewicz-Nilsson, J.; Wendt, A.; Andersson, N.; Hellstrand, P.; Grande, P.O.; et al. Deletion of the G protein-coupled receptor 30 impairs glucose tolerance, reduces bone growth, increases blood pressure, and eliminates estradiol-stimulated insulin release in female mice. *Endocrinology* **2009**, *150*, 687–698. [CrossRef]

70. Isensee, J.; Meoli, L.; Zazzu, V.; Nabzdyk, C.; Witt, H.; Soewarto, D.; Effertz, K.; Fuchs, H.; Gailus-Durner, V.; Busch, D.; et al. Expression pattern of G protein-coupled receptor 30 in LacZ reporter mice. *Endocrinology* **2009**, *150*, 1722–1730. [CrossRef]

71. Wang, C.; Dehghani, B.; Magrisso, I.J.; Rick, E.A.; Bonhomme, E.; Cody, D.B.; Elenich, L.A.; Subramanian, S.; Murphy, S.J.; Kelly, M.J.; et al. GPR30 contributes to estrogen-induced thymic atrophy. *Mol. Endocrinol.* **2008**, *22*, 636–648. [CrossRef]

72. Prossnitz, E.R.; Barton, M. The G-protein-coupled estrogen receptor GPER in health and disease. *Nat. Rev. Endocrinol.* **2011**, *7*, 715–726. [CrossRef]

73. Yu, T.; Yang, G.; Hou, Y.; Tang, X.; Wu, C.; Wu, X.A.; Guo, L.; Zhu, Q.; Luo, H.; Du, Y.E.; et al. Cytoplasmic GPER translocation in cancer-associated fibroblasts mediates cAMP/PKA/CREB/glycolytic axis to confer tumor cells with multidrug resistance. *Oncogene* **2017**, *36*, 2131–2145. [CrossRef] [PubMed]

74. Bonuccelli, G.; Whitaker-Menezes, D.; Castello-Cros, R.; Pavlides, S.; Pestell, R.G.; Fatatis, A.; Witkiewicz, A.K.; Vander Heiden, M.G.; Migneco, G.; Chiavarina, B.; et al. The reverse Warburg effect: Glycolysis inhibitors prevent the tumor promoting effects of caveolin-1 deficient cancer associated fibroblasts. *Cell Cycle* **2010**, *9*, 1960–1971. [CrossRef] [PubMed]

75. Whitaker-Menezes, D.; Martinez-Outschoorn, U.E.; Lin, Z.; Ertel, A.; Flomenberg, N.; Witkiewicz, A.K.; Birbe, R.C.; Howell, A.; Pavlides, S.; Gandara, R.; et al. Evidence for a stromal-epithelial "lactate shuttle" in human tumors: MCT4 is a marker of oxidative stress in cancer-associated fibroblasts. *Cell Cycle* **2011**, *10*, 1772–1783. [CrossRef] [PubMed]

76. Lappano, R.; Maggiolini, M. GPER is involved in the functional liaison between breast tumor cells and cancer-associated fibroblasts (CAFs). *J. Steroid Biochem. Mol. Biol.* **2018**, *176*, 49–56. [CrossRef] [PubMed]

77. Mizukami, Y. In vivo functions of GPR30/GPER-1, a membrane receptor for estrogen: From discovery to functions in vivo. *Endocr. J.* **2010**, *57*, 101–107. [CrossRef] [PubMed]

78. Allred, D.C.; Harvey, J.M.; Berardo, M.; Clark, G.M. Prognostic and predictive factors in breast cancer by immunohistochemical analysis. *Mod. Pathol.* **1998**, *11*, 155–168.

79. Bilancio, A.; Bontempo, P.; Di Donato, M.; Conte, M.; Giovannelli, P.; Altucci, L.; Migliaccio, A.; Castoria, G. Bisphenol A induces cell cycle arrest in primary and prostate cancer cells through EGFR/ERK/p53 signaling pathway activation. *Oncotarget* **2017**, *8*, 115620–115631. [CrossRef]

80. Zhou, X.; Wang, S.; Wang, Z.; Feng, X.; Liu, P.; Lv, X.B.; Li, F.; Yu, F.X.; Sun, Y.; Yuan, H.; et al. Estrogen regulates Hippo signaling via GPER in breast cancer. *J. Clin. Investig.* **2015**, *125*, 2123–2135. [CrossRef]

81. Pupo, M.; Pisano, A.; Abonante, S.; Maggiolini, M.; Musti, A.M. GPER activates Notch signaling in breast cancer cells and cancer-associated fibroblasts (CAFs). *Int. J. Biochem. Cell Biol.* **2014**, *46*, 56–67. [CrossRef]

82. Yuan, J.; Liu, M.; Yang, L.; Tu, G.; Zhu, Q.; Chen, M.; Cheng, H.; Luo, H.; Fu, W.; Li, Z.; et al. Acquisition of epithelial-mesenchymal transition phenotype in the tamoxifen-resistant breast cancer cell: A new role for G protein-coupled estrogen receptor in mediating tamoxifen resistance through cancer-associated fibroblast-derived fibronectin and β1-integrin signaling pathway in tumor cells. *Breast Cancer Res.* **2015**, *17*, 69.

83. Pandey, D.P.; Lappano, R.; Albanito, L.; Madeo, A.; Maggiolini, M.; Picard, D. Estrogenic GPR30 signalling induces proliferation and migration of breast cancer cells through CTGF. *EMBO J.* **2009**, *28*, 523–532. [CrossRef] [PubMed]

84. Tao, S.; He, H.; Chen, Q.; Yue, W. GPER mediated estradiol reduces miR-148a to promote HLA-G expression in breast cancer. *Biochem. Biophys. Res. Commun.* **2014**, *451*, 74–78. [CrossRef] [PubMed]

85. Vivacqua, A.; De Marco, P.; Santolla, M.F.; Cirillo, F.; Pellegrino, M.; Panno, M.L.; Abonante, S.; Maggiolini, M. Estrogenic GPER signaling regulates miR144 expression in cancer cells and cancer-associated fibroblasts (CAFs). *Oncotarget* **2015**, *6*, 16573–16587. [CrossRef] [PubMed]

86. Meijer, D.; van Agthoven, T.; Bosma, P.T.; Nooter, K.; Dorssers, L.C. Functional screen for genes responsible for tamoxifen resistance in human breast cancer cells. *Mol. Cancer Res.* **2006**, *4*, 379–386. [CrossRef] [PubMed]

87. van Agthoven, T.; Sieuwerts, A.M.; Meijer-van Gelder, M.E.; Look, M.P.; Smid, M.; Veldscholte, J.; Sleijfer, S.; Foekens, J.A.; Dorssers, L.C. Relevance of breast cancer antiestrogen resistance genes in human breast cancer progression and tamoxifen resistance. *J. Clin. Oncol.* **2009**, *27*, 542–549. [CrossRef] [PubMed]

88. De Marco, P.; Bartella, V.; Vivacqua, A.; Lappano, R.; Santolla, M.F.; Morcavallo, A.; Pezzi, V.; Belfiore, A.; Maggiolini, M. Insulin-like growth factor-I regulates GPER expression and function in cancer cells. *Oncogene* **2013**, *32*, 678–688. [CrossRef]

89. Recchia, A.G.; De Francesco, E.M.; Vivacqua, A.; Sisci, D.; Panno, M.L.; Andò, S.; Maggiolini, M. The G protein-coupled receptor 30 is up-regulated by hypoxia-inducible factor-1alpha (HIF-1alpha) in breast cancer cells and cardiomyocytes. *J. Biol. Chem.* **2011**, *286*, 10773–10782. [CrossRef]

90. Manjegowda, M.C.; Gupta, P.S.; Limaye, A.M. Hyper-methylation of the upstream CpG island shore is a likely mechanism of GPER1 silencing in breast cancer cells. *Gene* **2017**, *614*, 65–73. [CrossRef]

91. Giovannelli, P.; Di Donato, M.; Giraldi, T.; Migliaccio, A.; Castoria, G.; Auricchio, F. Targeting rapid action of sex steroid receptors in breast and prostate cancers. *Front. Biosci.* **2011**, *16*, 2224–2232. [CrossRef]

92. Cato, A.C.; Nestl, A.; Mink, S. Rapid actions of steroid receptors in cellular signaling pathways. *Sci. STKE* **2002**, *138*, re9. [CrossRef]

93. Prossnitz, E.R. New developments in the rapid actions of steroids and their receptors. *J. Steroid Biochem. Mol. Biol.* **2018**, *176*, 1–3. [CrossRef] [PubMed]

94. Lucas, T.F.; Siu, E.R.; Esteves, C.A.; Monteiro, H.P.; Oliveira, C.A.; Porto, C.S.; Lazari, M.F. 17 beta-estradiol induces the translocation of the estrogen receptors ESR1 and ESR2 to the cell membrane, MAPK3/1 phosphorylation and proliferation of cultured immature rat Sertoli cells. *Biol. Reprod.* **2008**, *78*, 101–114. [CrossRef] [PubMed]

95. Barton, M.; Filardo, E.J.; Lolait, S.J.; Thomas, P.; Maggiolini, M.; Prossnitz, E.R. Twenty years of the G protein-coupled estrogen receptor GPER: Historical and personal perspectives. *J. Steroid Biochem. Mol. Biol.* **2018**, *176*, 4–15. [CrossRef] [PubMed]

International Journal of
Molecular Sciences

Article

Sex Hormone Receptors in Benign and Malignant Salivary Gland Tumors: Prognostic and Predictive Role

Gabriella Aquino [1], Francesca Collina [1], Rocco Sabatino [1], Margherita Cerrone [1], Francesco Longo [2], Franco Ionna [2], Nunzia Simona Losito [1], Rossella De Cecio [1], Monica Cantile [1,*], Giuseppe Pannone [3] and Gerardo Botti [1]

[1] Pathology Unit, Istituto Nazionale per lo Studio e la Cura dei Tumori, "Fondazione G. Pascale", IRCCS, Naples 80131, Italy; gabryaquino@gmail.com (G.A.); francescacollina84@gmail.com (F.C.); roc.sabatino@gmail.com (R.S.); mcerrone1@virgilio.it (M.C.); s.losito@istitutotumori.na.it (N.S.L.); r.dececio@istitutotumori.na.it (R.D.C.); g.botti@istitutotumori.na.it (G.B.)

[2] Head and Neck Surgery Unit, Istituto Nazionale per lo Studio e la Cura dei Tumori, "Fondazione G. Pascale", IRCCS, Naples, 80131, Italy; f.longo@istitutotumori.na.it (F.L.); f.ionna@istitutotumori.na.it (F.I.)

[3] Department of Clinical and Experimental Medicine, Pathological Anatomy Unit—University of Foggia, Foggia 71100, Italy; giuseppepannone@virgilio.it

* Correspondence: m.cantile@istitutotumori.na.it; Tel.: +39-081-590-3745; Fax: +39-081-590-3718

Received: 21 November 2017; Accepted: 19 January 2018; Published: 30 January 2018

Abstract: The role of sex hormone receptors in human cancer development and progression has been well documented in numerous studies, as has the success of sex hormone antagonists in the biological therapy of many human tumors. In salivary gland tumors (SGTs), little and conflicting information about the role of the estrogen receptor alpha (ERα), progesterone receptor (PgR) and androgen receptor (AR) has been described and in most cases the use of sex hormone antagonists is not contemplated in clinical practice. In this study, we analyzed a panel of sex hormone receptors that have not been widely investigated in SGTs—ERα, PgR, AR, but also ERβ and GPR30—to define their expression pattern and their prognostic and predictive value in a case series of 69 benign and malignant SGTs. We showed the aberrant expression of AR in mucoepidermoid and oncocytic carcinoma, a strong relation between cytoplasmic ERβ expression and tumor grade, and a strong correlation between nuclear GPR30 expression and disease-free survival (DFS) of SGT patients.

Keywords: sex hormone receptors; salivary gland tumors; therapeutic targets

---

1. Introduction

Salivary gland tumors (SGT) are rare tumors, representing approximately 0.5% of all human cancers and less than 5% of head and neck lesions [1]. The WHO classification identifies 24 different malignant subtypes with different clinical courses and variable prognoses, mainly represented by primary epithelial tumors that account for approximately 88% of the SGTs [2]. Mucoepidermoid tumor (MEC), Salivary Duct Carcinoma (SDC) and adenoid cystic carcinoma (AdCC) represent the most frequent and often the more aggressive lesions [1]. Until today, surgical excision represents the only choice of treatment, with radio and/or chemotherapy in case of advanced disease and loco-regional recurrences. The application of new therapeutic strategies that are mainly based on the employment of biological drugs should be integrated into the management of these patients.

The overexpression of several sex hormone receptors, in particular, estrogen receptor alpha (ERα), progesterone receptor (PgR), and androgen receptor (AR), suggests their fundamental role in tumor pathogenesis and progression [3–7]. The production of sex hormone antagonists and their success in the treatment of patients with ERα+ and PgR+ breast carcinomas and AR+ prostate carcinomas have also suggested the investigation of the expression of these receptors in other tumors including SGTs [8].

Sex steroid hormones appear to play the main role in the physiology of the human oral cavity and salivary glands. However, most of the studies focused on the expression of ERα and PR report conflicting results. Alternatively, the expression and role of AR in SGTs are well documented [9,10]. The other estrogen receptor (ERβ) was described in salivary gland adenocarcinoma cell lines and certain salivary gland carcinomas such as AdCC and Pleomorphic adenoma (PA) [11,12]. The structure of ERβ is homologous to that of ERα and its DNA-binding domain is 96% conserved compared to ERα, suggesting that ERβ could bind the same target genes [13]. Specific ERβ isoforms are able to activate specific signal transduction pathways starting from the cytoplasm or plasma membrane, which may explain the effect of E2 in the modulation of cytoskeletal remodeling and the migration of salivary gland adenocarcinoma cells [14].

Whereas ERα and ERβ mediate the genomic estrogen signaling, the third membrane-bound Estrogen Receptor GPR30 (GPER) mediates the non-genomic signaling mechanisms. Several studies reported that the ligand activation of GPR30 signaling, coupled with the upregulation of specific GPER genes, was involved in the proliferation of tumor cells, suggesting that GPER can contribute to tumorigenesis [15,16]. On the role of GPR30 in SGTs, only one study showed GPR30 expression in oral epithelia like salivary glands and tongue [17].

Overall, little information is reported in the literature on the role of ERβ and GPR30 in SGTs. In this study, we aimed to analyze a panel of sex hormone receptors, such as ERα, ERβ, GPR30, PgR, and AR, in a case series of 75 SGTs of different histotypes to better define their expression pattern and their prognostic and predictive value in these tumors.

2. Results

2.1. Characteristics of SGTs Patients

In the study, only the patients with a complete panel of clinical-pathological features have been included, while Kaplan–Meier analysis has been carried out on selected patients with clinical outcome. The patients initially selected were 69 in number, and their samples have been included in Tissue Micro Array (TMA), however, the number of samples evaluable for statistical elaboration ranges from 54 to 62 cases, because of skipping cores for the different markers. The SGTs TMA was built with 36 cases of benign tumors (pleomorphic adenoma (PA), myoepithelioma, basal cell adenoma, Warthin tumor, and oncocytoma) and 33 cases of malignant tumors (MEC, acinic cell carcinoma (ACC), adenocarcinoma, mixed tumor, carcinoma ex pleomorphic adenoma (Ca ex PA), AdCC, oncocytic carcinoma and salivary duct carcinoma (SDC)). The prevalent location of these lesions is the parotid gland. All clinical pathological information of patients is summarized in Table 1.

Table 1. Main Clinical-Pathological data.

Patient Features	Number of Patients	69
	Median Age (Range)	60 (17–87) Years
Sex	Male	41 (59.4%)
	Female	28 (40.6%)
Lesion	Benign	36 (52.2%)
	Malign	33 (47.8%)
Site	Parotid	59 (85.5%)
	SG	10 (14.5%)
Grading	G1	14 (42.4%)
	G2/G3	19 (57.6%)
	Benign (without grading)	36
Ki67 Score	≤5%	42 (60.9%)
	>5%	20 (29%)
	NA	7 (10.1%)
Cell Type Differentiation	Epithelial	38 (55.1%)
	Myoepithelial	7 (10.1%)
	Mixed	24 (34.8%)

Table 1. *Cont.*

Patient Features	Number of Patients Median Age (Range)	69 60 (17–87) Years
	MEC	13 (18.8%)
	ACC	9 (13%)
	CA ex PA	3 (4.3%)
	Adenocarcinoma	2 (2.9%)
	AdCC	1 (1.4%)
	SDC	1 (1.4%)
Histotype	Oncocytic CA	1 (1.4%)
	Mixed tumor	3 (4.3%)
	PA	18 (26.1%)
	Warthin's tumors	9 (13%)
	Myoepithelioma	7 (10.1%)
	Oncocytoma	1 (1.4%)
	Basal cell adenoma	1 (1.4%)

SG: Salivary Galnd; G1: Grading 1 G2: Grading 2 G3: Grading 3; MEC: Mucoepidermoid carcinoma; ACC: Acinic cell carcinoma; CA: Carcinoma; PA: Pleomorphic adenoma; AdCC: Adenoid cystic carcinoma SDC: Salivary ductal carcinoma.

2.2. Immunohistochemical Expression of AR, ERβ and GPR30, and Relation with Clinical-Pathological Features and Survival in SGTs

Little and often conflicting information about the role of sex hormone receptors in SGTs has been provided and, consequently, the use of specific biological drugs is not usually planned for these tumor diseases. For this reason, we analyzed a panel of sex hormone receptors in a case series of patients with benign and malignant SGTs.

For all biomarkers, we considered both nuclear and cytoplasmic staining. Receptors ERα and PgR are never expressed in our series, in line with the literature [18]. In detail, we detected only nuclear AR expression in 15/61 (24%) of SGT samples in both malignant and benign SGT lesions. A total of eight cases were not considered evaluable.

Considering the stratification of the lesions based on their cell differentiation, we detected AR expression in 17% of epithelial SGTs, in 28% of myoepithelial lesions and in 35% of mixed SGTs. (Figure 1A).

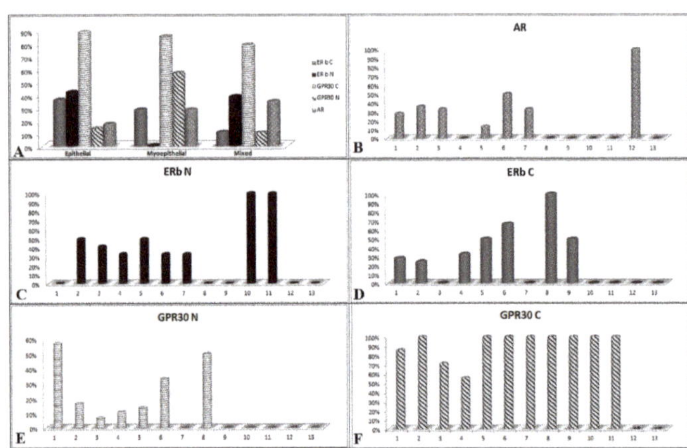

Figure 1. Schematic representation of distribution of Androgen Receptor (AR), Estrogen Receptor Beta (ERβ) and G protein-coupled receptor 30 (GPR30 IHC) expression in salivary gland tumors (SGTs): (**A**) AR, ERβ and GPR30 expression in cell differentiation SGT types (epithelial, myoepithelial and mixed); (**B**) nuclear AR expression in different histotypes; (**C**) nuclear ERβ expression in different SGT histotypes; (**D**) cytoplasmic ERβ expression in different SGT histotypes; (**E**) nuclear GPR30 expression in different SGT histotypes; (**F**) cytoplasmic GPR30 expression in different SGT histotypes. *X* = SGTs histotypes; *Y* = number of positive samples in percentage terms.

In the context of benign lesions, AR was mainly expressed in PA (33%) and in 28% of myoepithelioma samples, and interestingly in 36% of MEC, in sporadic cases of Ca ex PA, and oncocytic carcinoma (Figures 1B and 2). Their aberrant expression in malignant SGTs was sporadically reported in MEC, a very poor prognosis tumor, and never reported for oncocytic carcinoma, suggesting the use of AR antagonists in therapeutic strategies for these patients.

Regarding ERβ we detected nuclear expression in 21/58 (36%) of SGTs and cytoplasmic staining in 16/58 (27%) of SGTs. We never detected nuclear and cytoplasmic ERβ co-expression. A total of 11 cases were not considered evaluable.

Regarding cell differentiation types, we detected cytoplasmic expression of ERβ in 36% of epithelial SGTs, in 28% of myoepithelial lesions, and in 11% of mixed SGTs. Nuclear expression was present in 42% epithelial SGTs, in 38% of mixed SGTs, and never detected in myoepithelial lesions (Figure 1A). In detail, nuclear ERβ was present in 40% of malignant lesions, mainly in 50% of MEC samples and in 33% of ACC. Moreover, we detected ERβ nuclear expression in 25% of benign lesions, mainly represented by PA and Warthin tumor (Figures 1C and 3). Cytoplasmic expression of ERβ was detected in 33% of malignant lesions, mainly in 25% of MEC samples and in 33% of ACC, followed by sporadic cases of mixed tumors and adenocarcinoma. Moreover, we detected ERβ cytoplasmic expression in 19% of benign lesions, above all in myoepithelioma and Warthin tumors (Figures 1D and 4). Also, in this case, the overexpression of ERβ in several malignant SGTs can suggest the use of antagonists of estrogen receptors, with equivalent affinities for ERβ and ERα [19], in these tumor patients.

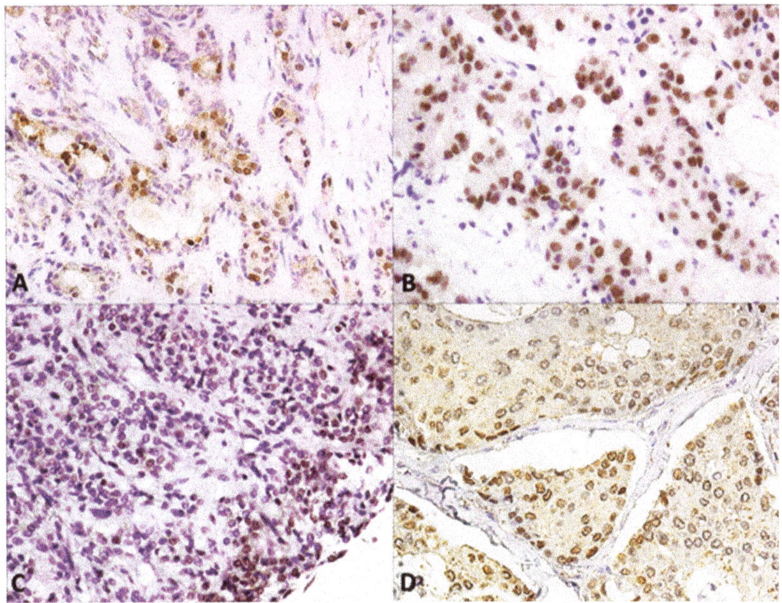

Figure 2. Nuclear AR IHC staining of SGTs samples: (**A**) Pleomorphic Adenoma (PA); (**B**) Oncocytic carcinoma; (**C**) Myoepithelioma; (**D**) Mucoepidermoid Carcinoma (MEC) (Magnification 20×).

Finally, we detected cytoplasmic staining of GPR30 in 34/62 (86%) of specimens with nuclear co-expression in 11/62 (18%) of SGTs. A total of 7 cases were not considered evaluable.

Regarding cell differentiation types we detected cytoplasmic expression of GPR30 in 88% of epithelial SGTs, in 85% of myoepithelial lesions and in 78% of mixed SGTs. Nuclear staining was detected respectively in 13%, 57% and 10% of epithelial, myoepithelial and mixed SGTs (Figure 1A).

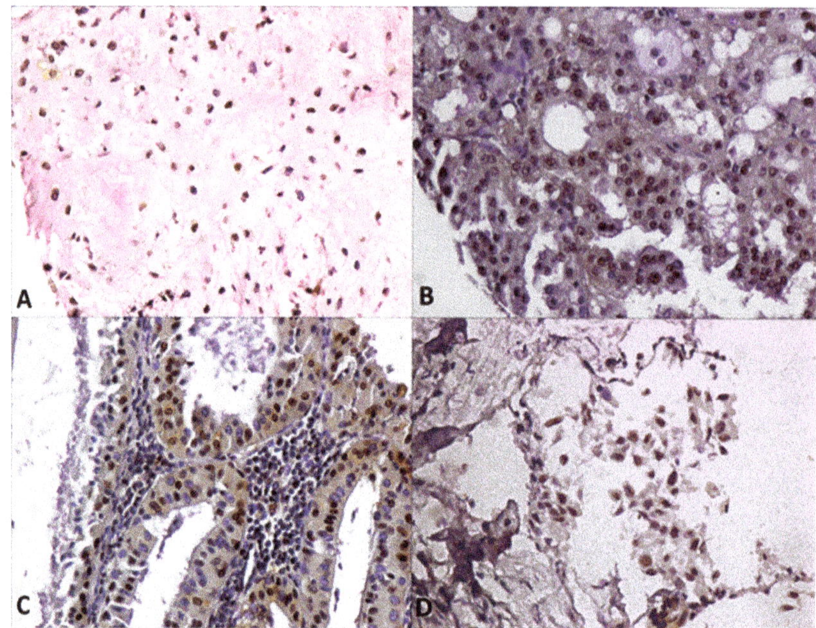

Figure 3. Nuclear ERβ IHC staining of SGTs samples: (**A**) PA; (**B**) acinic cell carcinoma (ACC); (**C**) Warthin's tumor; (**D**) MEC (Magnification 20×).

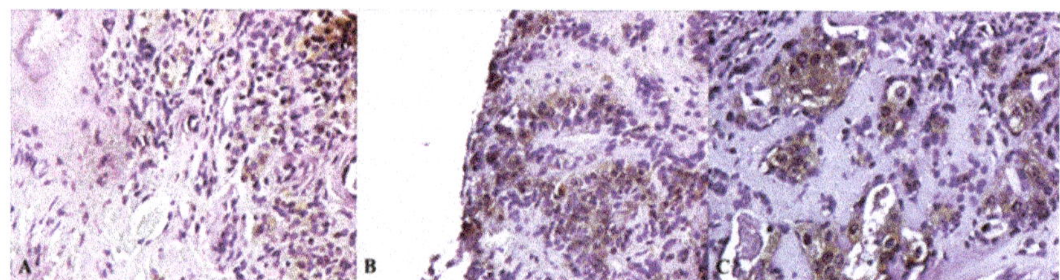

Figure 4. Cytoplasmic ERβ IHC staining of SGTs samples: (**A**) ACC; (**B**) Myoepithelioma; (**C**) MEC (Magnification 20×).

Cytoplasmic GPR30 expression was present in all cases of MEC and in most of other malignant lesions. In benign SGTs its expression was prevalent in myoepithelioma and PA samples. (Figure 5).

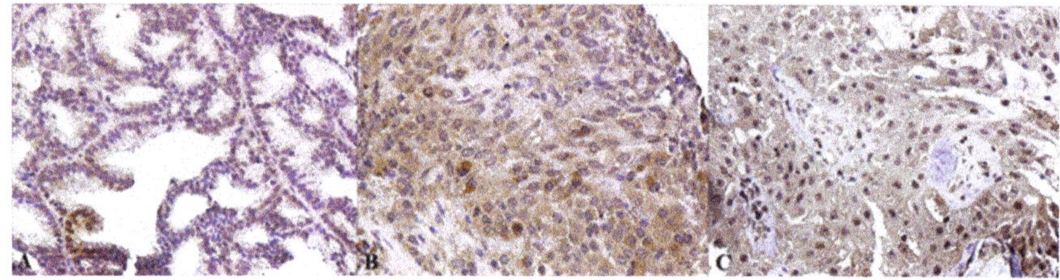

Figure 5. Nuclear GPR30 IHC staining of SGTs samples: (**A**) ACC; (**B**) Myoepithelioma; (**C**) MEC (Magnification 20×).

The nuclear GPR30 positivity was detected in 57% of myoepithelioma and in sporadic cases of PA and Warthin's tumors. In malignant SGTs we detected nuclear GPR30 in 16% of MECs and in sporadic cases of ACCs, and adenocarcinoma (Figure 6).

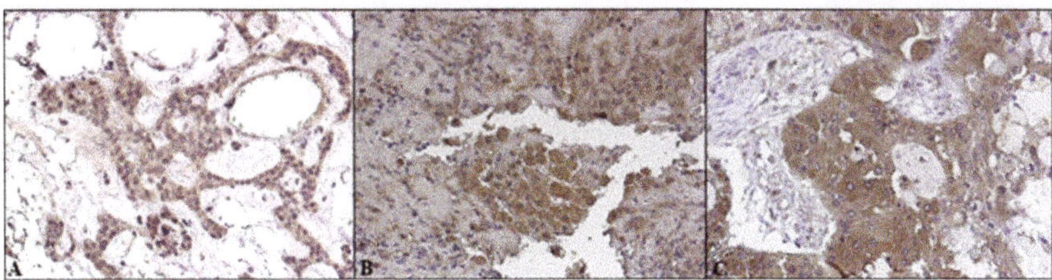

Figure 6. Cytoplasmic GPR30 IHC staining of SGTs samples: (**A**) PA; (**B**) Myoepithelioma; (**C**) MEC (Magnification 20×).

For the statistical elaboration, we considered tumor grade only in malignant tumors. Based on statistical elaboration of nuclear AR expression with the clinical-pathological features of SGTs, we showed no statistical significance with age, gender, site location, grade, cell differentiation, and proliferation index (Table 2). Cytoplasmic ERβ expression was significantly associated only with tumor grade (p-value = 0.052), while no statistical association with clinical-pathological features exist for nuclear ERβ expression. Similarly, no statistical association with clinical-pathological characteristics exist for cytoplasmic and nuclear GPR30 expression, except a trend of statistical association between cytoplasmic GPR30 expression and tumor grade (p-value = 0.087). All data are schematized in Table 2.

Table 2. Statistical association of AR, ERβ and GPR30 tumor expression with clinical pathological features of SGT patients. (SG = Submandibular Gland)

Patient Features		Nuclear AR				Cytoplasmic ERβ				Nuclear ERβ				Cytoplasmic GPR30				Nuclear GPR30			
		Negative	Positive	p-Value	R Pearson	Negative	Positive	p-Value	R Pearson	Negative	Positive	p-Value	R Pearson	Negative	Positive	p Value	R Pearson	Negative	Positive	p-Value	R Pearson
Age	≤60	24	10	0.326	−0.126	25	6	0.133	0.197	18	13	0.331	−0.128	3	29	0.235	−0.151	24	8	0.122	−0.196
	>60	22	5			17	10			19	8			6	24			27	3		
Sex	Male	27	8	0.715	0.047	25	10	0.836	−0.027	22	13	0.855	−0.024	3	32	0.130	−0.192	30	5	0.417	0.103
	Female	19	7			17	6			15	8			6	21			21	6		
Site	Parotid	41	13	0.795	0.033	35	14	0.695	−0.051	32	17	0.576	0.073	8	45	0.754	0.040	42	11	0.132	−0.191
	SG	5	2			7	2			5	4			1	8			9	0		
Lesion	Benign	22	9	0.413	−0.105	21	6	0.394	0.112	18	9	0.671	0.056	4	27	0.718	−0.046	24	7	0.319	−0.127
	Malignant	24	6			21	10			19	12			5	26			27	4		
Grade	G1	10	2	0.709	0.068	12	2	0.052	0.349	9	5	0.756	0.056	4	10	0.087	0.307	13	1	0.385	0.156
	G2–G3	14	4			9	8			10	7			1	16			14	3		
Ki67	≤5%	29	9	0.904	−0.016	26	9	0.392	0.116	23	12	0.570	0.077	8	31	0.150	0.191	32	7	0.704	0.050
	>5%	14	4			12	7			11	8			1	17			14	4		

Regarding the relation with clinical outcome of SGT patients, Kaplan–Meier curves referred to DFS and OS are illustrated in Figures 7 and 8. We showed no statistical association with DFS and OS for both AR and nuclear and cytoplasmic expression of ERβ. Regarding GPR30 we showed a strong statistical significance between its nuclear expression and DFS (p-value = 0.055) (Figure 8D). The relationship between nuclear GPR30 and DFS highlighted the never reported prognostic role of this marker in SGTs.

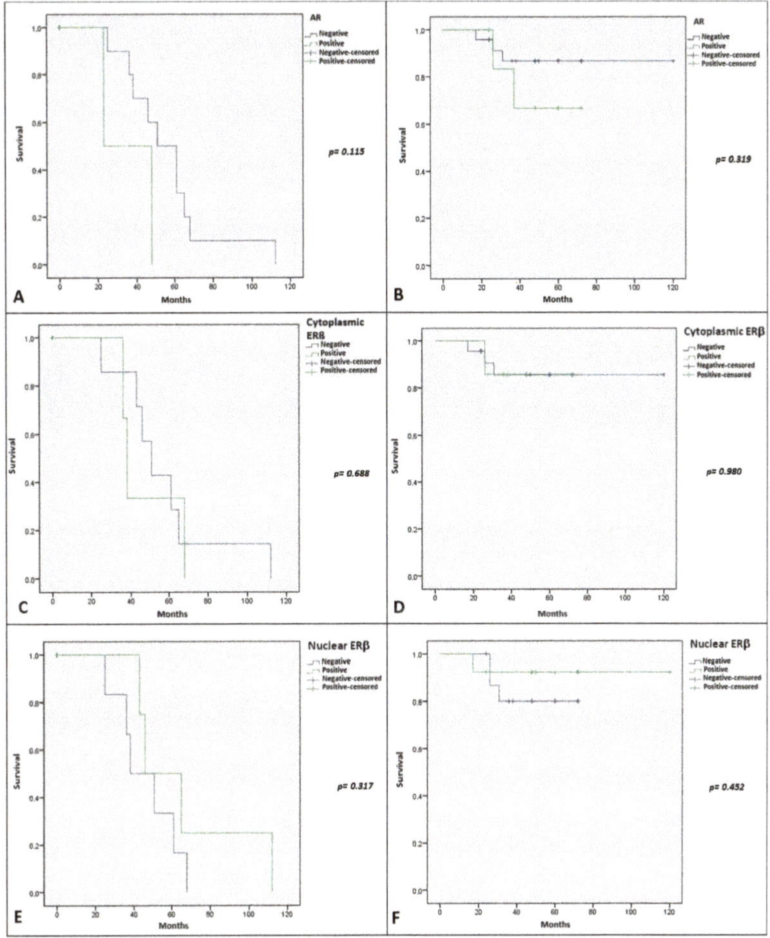

Figure 7. (**A**)Kaplan–Meier plot for disease-free survival (DFS) in patients with SGT stratified by AR IHC expression. The green line represents patients with AR nuclear positivity; (**B**) Kaplan–Meier plot for Overall survival (OS) in patients with SGT stratified by AR IHC expression. The green line represents patients with AR nuclear positivity; (**C**) Kaplan–Meier plot for DFS in patients with SGT stratified by cytoplasmic ERβ IHC expression. The green line represents patients with cytoplasmic ERβ positivity; (**D**) Kaplan–Meier plot for OS in patients with SGT stratified by Cytoplasmic ERβ IHC expression. The green line represents patients with cytoplasmic ERβ positivity; (**E**) Kaplan–Meier plot for DFS in patients with SGT stratified by nuclear ERβ IHC expression. The green line represents patients with nuclear ERβ positivity; (**F**) Kaplan–Meier plot for OS in patients with SGT stratified by nuclear ERβ *IHC* expression level. The green line represents patients with nuclear ERβ positivity.

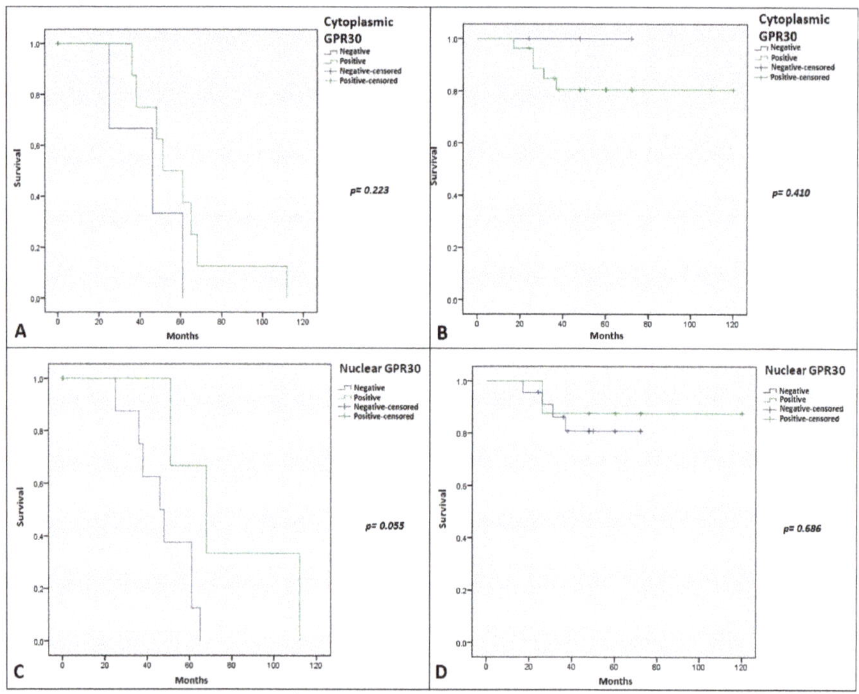

Figure 8. (**A**) Kaplan–Meier plot for disease-free survival (DFS) in patients with SGT stratified by Cytoplasmic GPR30 IHC expression. The green line represents patients with Cytoplasmic GPR30 positivity; (**B**) Kaplan–Meier plot for overall survival (OS) in patients with SGT stratified by GPR30 IHC expression. The green line represents patients with Cytoplasmic GPR30 positivity; (**C**) Kaplan–Meier plot for DFS in patients with SGT stratified by Nuclear GPR30 IHC expression (*p*-value = 0.055). The green line represents patients with Nuclear GPR30positivity; (**D**) Kaplan–Meier plot for OS in patients with SGT stratified by Nuclear GPR30 IHC expression. The green line represents patients with Nuclear GPR30 positivity.

3. Discussion

In recent years, many studies have focused on the expression of sex hormone receptors in human cancer and on the mechanisms through exerting their actions and influence the progression of tumor diseases. Moreover, the development of sex hormone antagonists and their successful employment in biological therapies for several tumors has suggested the evaluation of their expression and/or activity in different cancer types. However, in SGTs there is fragmentary and often conflicting information about the role of sex hormone receptors, and, for this reason, the use of biological drugs is not contemplated in clinical practice in the majority of the cases.

In our study, we analyzed a case series of patients with benign and malignant SGTs included in a TMA and correlated their expression with clinical-pathological parameters and outcomes.

In our SGT case series, we have never detected the expression of ERα and PgR. In literature, whereas benign salivary glands tumors were negative for hormone receptors expression [20], widely disparate results about ERα and PR expression in various malignant SGTs have been reported. Early studies showed immuno-positivity of ERα in 8% of SDC, with a total absence of PgR expression [21], a marked expression of PgR and absence of ERα expression in AdCC [22,23], while sporadic cases of ACC and MEC showed a positivity for both receptors [24]. Another study has described the absence of ERα expression both in AdCC and MEC [25]. More recent studies described ER and PgR positivity in only a few cases [20] while, as in our case, a large case series (139 salivary glands tumors) study never detected ERα and PgR positivity [18].

Regarding AR expression, we detected nuclear AR expression mainly in several benign lesions such as PA and myoepithelioma, but the aberrant AR expression was also identified in several malignant lesions. Whereas AR expression was abundantly documented in Ca ex PA, our data also showed the interesting expression in many cases of MEC and oncocytic carcinoma.

A rich literature documented the expression and the role of AR in salivary glands tumors. Early studies described a very high IHC AR expression in SDC [9,10] with a more significant expression in men (79%) than in women (33%) [26]. Little information is available in literature about the role of AR in other SGTs. The absence of its expression was reported in AdCC, MEC, and ACC [27]. In PA a focal immunohistochemical expression of AR was described [28], while its expression was detected in 90% of Ca ex PA, suggesting an AR role in malignant tumor evolution [29].

Concerning the therapeutic potential of anti-AR drugs, several studies reported the benefits of anti-androgen therapy, in particular in the SDC histotype. In a series of 10 patients with an overexpression of AR, 50% of them was enormously benefited from treatment with bicalutamide [30]. Our data, in particular the aberrant expression of AR in several MEC and oncocytic carcinoma could suggest the potential use of anti-AR drugs also in these tumor types.

Regarding ERβ expression, we detected its positivity both at nuclear and cytoplasmic level with a prevalent expression in epithelial malignant lesions such as MEC samples and ACC, while myoepithelial lesions never present ERβ nuclear expression.

Expression of ERβ was reported at high levels in oral tissues, mainly in keratinocytes and salivary gland acinar and ductal cells [31]. Overexpression of ER-β was described in four cases of pediatric MEC and in ACC cell line [32], while nuclear overexpression of ER-β was detected also in 71% of ACC FPPE tissues, with the average expressions higher in women, and in the cases with a cribriform architecture [11]. ERβ was also detected in several cases of PA of the salivary gland [12].

Several studies showed that antagonists of estrogen receptors, can have therapeutic effects in preclinical models, in particular in ERβ+ TNBC models. Oral estradiol, approved for treatment of metastatic breast cancer has equivalent affinities for ERβ and ERα [19]. In fact, ERβ can bind other ligands with rather higher affinity than ERα, such as 4-hydroxytamoxifen, the phytoestrogen genistein, and, testosterone derivatives, 3βAdiol [33]. This suggests the possibility of its use to target ERβ in TNBC [29] but also in other ERα+ tumors. Moreover, several studies showed that higher ERβ expression was an independent predictor of better tamoxifen response [34,35] and overexpression of ERβ1 was also associated with increased sensitivity to 4-hydroxytamoxifen [36].

In our SGTs case series, while the nuclear ERβ expression does not appear to be associated with clinical outcomes, cytoplasmic ERβ staining showed a strong association with tumor grade, highlighting its strong prognostic value. It was reported that different ERβ variant isoforms can be localized in the cytoplasm and plasma membrane, showing variable expression in cancer tissues and influencing cancer progression and response to therapy [37]. Our results suggest that cytoplasmic ERβ signaling in SGTs may be more important for patient outcome than its nuclear signaling. This is probably due to ERβ2 isoform which is already documented to be strongly related to poorer prognosis in breast cancer [38]. Several studies showed the same findings in other tumor types, such as ovarian cancer, squamous cell carcinoma [39,40]. For these tumors, the use of estrogen receptor antagonists could be suggested in clinical practice.

Only one study in the literature reported the expression of GPR30 in oral tissues [20]. GPR30 (GPER), as a 7-transmembrane GPCR and is predominantly, though not exclusively, localized on intracellular membranes, particularly on those of the endoplasmic reticulum and Golgi apparatus [41] in several tissues such as reproductive tissues, heart, intestines, ovary, CNS, pancreatic islets, adipose tissue, skeletal muscle, liver, neurons, and inflammatory cells [42].

We detected its cytoplasmic staining in most SGT specimens, particularly in MEC. Furthermore, its nuclear staining was prevalent in several benign lesions but also in a discrete number of MEC and ACC. Cheng et al. demonstrated that retrograde transport of GPR30 from the plasma membrane towards the nucleus occurs with a consecutive accumulation of GPR30 in the perinuclear space followed by a later dispersion in the cytoplasm [43]. Recent studies showed that the different location of GPR30—cytoplasmic and nuclear locations—can reflect distinct tumor properties in breast cancer [44], and the lack of GPR30 expression in the plasma membrane can be associated with excellent long-term prognosis in ERα and PgR-positive tamoxifen-treated primary breast cancer [45]. This trend reflects our data. In fact, in our series nuclear expression of GPR30, it was statistically associated with a better DFS in SGT patients. Although the subcellular GPR30 trafficking process (which is probably related to a functional receptor modulation) has never been described in SGTs, we can speculate a dynamic intracellular shift strongly related to SGT cancer progression.

A non-steroidal, high-affinity GPR30 agonist G-1 has been developed to dissect GPR30-mediated estrogen responses from those mediated by classic estrogen receptors [46]. Moreover, several highly selective GPR30 antagonists, such as G15 and G36, were identified [47]. In particular, G36 has a better activity compared to G15 in a range of functional assays, both in vitro and in vivo [48]. In an endometrial tumor cell model, G36 greatly reduces growth of estrogen-stimulated cells, suggesting that GPR30 may play a critical role in endometrial carcinogenesis and, therefore, providing G36 as a novel target for prognosis and treatment [49].

In conclusion, our data highlighted the aberrant expression of several sex hormone receptors, in particular of alternative estrogen receptors, such as ERβ and GPR30 in SGTs, showing their prognostic value and suggesting consideration of them as new biological targets.

4. Material and Methods

4.1. Patients with Salivary Glands Tumors

75 patients admitted to the National Cancer Institute "Giovanni Pascale" of Naples, between 2012 and 2017, were recruited in this study. All patients had provided written informed consent for the use of samples according to the institutional regulations and the study was approved by the ethics committee of the National Cancer Institute "Giovanni Pascale" and was registered "Bio-Banca Istituzionale BBI" Deliberation (NO. 15 del, 20 Jan. 2016).

All cases have been reviewed according to WHO 2017 classification criteria [2] using standard tissue sections. Clinic-pathological characteristics, including tumor–node–metastasis (TNM) stage, were collected. Medical records have been reviewed for clinical information, including histologic parameters assessed on standard H&E-stained slides.

4.2. TMA Building

A Prognostic-Tumor Array was built using 75 tumor tissue samples. H&E staining of 4 μm TMA section was used to verify all samples. One core from tumor areas of each subtype tumor was arrayed in a recipient block. All tumors and controls were reviewed by two experienced pathologists (Giuseppe Pannone and Nunzia Simona Losito). Discrepancies for the same case were resolved in a joint analysis. Tissue cylinders with a diameter of 1 mm were punched from morphologically representative tissue areas of each "donor" tissue block and brought into one recipient paraffin block (3 core of tissue × 1 mm) using a semi-automated tissue array (Galileo TMA CK 3500 Tissue Micro arrayer; ISE TMA Software, Integrated System Engineering, Milano, Italy).

4.3. Immunohistochemistry Analysis

Immunohistochemical staining was carried out on slides from formalin-fixed, paraffin embedded tissues (FPPE), in order to evaluate the expression of ERα, ERβ, GPR30, PgR, and AR. FPPE slides were de-paraffinized in xylene and rehydrated through graded alcohols. Antigen retrieval was performed with slides heated in 0.0.1 M citrate buffer (pH 6.0) in a bath for 20 min at 97 °C. After antigen retrieval, the slides were allowed to cool. The slides were rinsed with TBS and the endogenous peroxidase has inactivated with 3% hydrogen peroxide. After protein block (BSA 5% in PBS 1x), the slides were incubated with primary antibody to human ERα (Monoclonal Mouse Anti-Human ERα, Clone ID5, dilution 1:35, Dako North America, Inc., Carpinteria, CA, USA), PR (Monoclonal Mouse Anti-Human PR, Clone 636, dilution 1:50, Dako North America, Inc., Carpinteria, CA, USA), Ki67 (Monoclonal Mouse Anti-Human Ki67 Ag Clone MIB-1, dilution 1:75, Dako North America, Inc., Carpinteria, CA, USA) for 30 min, AR (monoclonal mouse anti-human AR antibody clone AR441, dilution 1:75, #M3562; Dako North America, Inc., Carpinteria, CA, USA), GPR30 (polyclonal rabbit antibody, clone sc-48524-R, dilution 1:300, Santa Cruz Biotechnology, Dallas, TX, USA) and ERβ (Monoclonal Mouse Anti-Human ERβ, clone PPG5/10, dilution 1:30, Dako North America, Inc., Carpinteria, CA, USA) overnight. Sections were incubated with mouse anti-rabbit or goat anti-mouse secondary IgG biotinylated secondary antibody for 30 min. Immunoreactivity was visualized by means of avidin–biotin–peroxidase complex kit reagents (Novocastra, Newcastle, UK) as the chromogenic substrate. Finally, sections were weakly counterstained with hematoxylin and mounted.

4.4. Evaluation of Immunostaining

Antigen expression was independently evaluated by two experienced pathologists (GP/SL) using light microscopy. All values of immunostaining were expressed in percentage terms of positive cells and intensity. The percentage of positive cancer cells was evaluated in each sample by counting the number of positive cells over the total cancer cells in 10 non-overlapping fields using $400\times$ magnification. The cutoff used to distinguish "positive" from "negative" cases was $\geq$1% ERα/PR positive tumor cells. For the proliferative index Ki67 was defined as the percentage of immuno-reactive tumor cells out of the total number of cells. The percentage of positive cells per case was scored according to 2 different groups: group 1: <5% (low proliferative activity); group 2: >5% (high proliferative activity). For nuclear AR expression the cutoff used to distinguish "positive" from "negative" cases was $\geq$1% AR-positive tumor cells. For ERβ expression was considered the percentage of positive cells for both nuclear and cytoplasmic staining. For GPR30, being positive in the most of cells for each sample, we considered the intensity of the reaction as negative, weak, intermediate, and strong (0, 1+, 2+, 3+) (Supplementary Table S1).

4.5. Statistical Analysis

The association between ERα, ERβ, GPR30, PgR and AR expression with clinical-pathological parameters and was conducted using the χ^2 and Student's *t*-test.

The Pearson χ^2 test was used to determine whether a relationship existed between the variables included in the study. The level of significance was defined as $p < 0.05$. Overall survival (OS) and disease-free survival (DFS) curves were calculated using the Kaplan–Meier method with significance valuated using the Mantel–Cox log-rank test. All the statistical analyses were carried out using the Statistical Package for Social Science v. 20 software (SPSS Inc., Chicago, IL, USA). OS was defined as the time from diagnosis (first biopsy) to death by any cause or until the most recent follow-up. DFS was measured as the time from diagnosis to the occurrence of progression, relapse after complete remission, or death from any cause. DFS had a value of zero for patients who did not achieve complete remission. The follow-up duration was five years.

Supplementary Materials: Supplementary materials can be found at http://www.mdpi.com/xxx/s1.

Acknowledgments: This study was supported by the Italian Ministry of Health and bio-materials (Tissues and blood) for in vivo studies were provided by Institutional BioBank (BBI) of INT Fondazione Pascale.

Author Contributions: Gabriella Aquino, Monica Cantile and Giuseppe Pannone were responsible for the conception and design of the study. Francesco Longo and Franco Ionna were responsible for provision of patients clinical information. Rocco Sabatino, Gabriella Aquino and Margherita Cerrone collected samples for immunohistochemical analysis. Nunzia Simona Losito, Rossella De Cecio, Giuseppe Pannone and Gerardo Botti were responsible for immunohistochemical evaluation. Francesca Collina was responsible for statistical analysis. Monica Cantile was responsible for paper elaboration and revision. All authors were involved in manuscript writing and provided final approval of the manuscript.

Conflicts of Interest: The authors declare no conflict of interest.

References

1. Cheu, W.; Chan, J.K. Salivary gland tumours. In *Diagnostic Histopathology of Tumours*, 2nd ed.; Fletcher, C.D., Ed.; Churchill Livingstone: London, UK, 2000; p. 231.
2. El-Naggar, A.K.; Chan, J.K.C.; Grandis, J.R.; Takata, T.; Slootweg, P.J. (Eds.) *WHO Classification of Head and Neck Tumours*, 4th ed.; IARC Press: Lyon, France, 2017; p. 160.
3. Rochefort, H.; Chalbos, D. The role of sex steroid receptors on lipogenesis in breast and prostate carcinogenesis: A viewpoint. *Horm. Cancer* **2010**, *1*, 63–70. [CrossRef] [PubMed]
4. Townsend, E.A.; Miller, V.M.; Prakash, Y.S. Sex differences and sex steroids in lung health and disease. *Endocr. Rev.* **2012**, *33*, 1–47. [CrossRef] [PubMed]
5. González-Arenas, A.; Agramonte-Hevia, J. Sex steroid hormone effects in normal and pathologic conditions in lung physiology. *Mini Rev. Med. Chem.* **2012**, *12*, 1055–1062. [CrossRef] [PubMed]
6. Cleve, A.; Fritzemeier, K.H.; Haendler, B.; Heinrich, N.; Möller, C.; Schwede, M.; Wintermantel, T. Pharmacology and clinical use of sex steroid hormone receptor modulators. *Handb. Exp. Pharmacol.* **2012**, *214*, 543–587.
7. Higa, G.M.; Fell, R.G. Sex Hormone Receptor Repertoire in Breast Cancer. *Int. J. Breast. Cancer* **2013**, *2013*, 284036. [CrossRef] [PubMed]
8. Giovannelli, P.; Di Donato, M.; Giraldi, T.; Auricchio, F. Targeting rapid action of sex steroid receptors in breast and prostate cancers. *Front Biosci.* **2011**, *16*, 2224–2232. [CrossRef]
9. Kapadia, S.B.; Barnes, L. Expression of androgen receptor, gross cystic disease fluid protein, and CD44 in salivary duct carcinoma. *Mod. Pathol.* **1998**, *11*, 1033–1038. [PubMed]
10. Fan, C.Y.; Melhem, M.F.; Hosal, A.S.; Grandis, J.R.; Barnes, E.L. Expression of androgen receptor, epidermal growth factor receptor, and transforming growth factor α in salivary duct carcinoma. *Arch. Otolaryngol.-Head Neck Surg.* **2001**, *127*, 1075–1079. [CrossRef] [PubMed]
11. Marques, Y.M.; Giudice, F.S.; Freitas, V.M.; Abreu e Lima Mdo, C.; Hunter, K.D.; Speight, P.M.; Machado de Sousa, S.C. Oestrogen receptor β in adenoid cystic carcinoma of salivary glands. *Histopathology* **2012**, *60*, 609–616. [CrossRef] [PubMed]
12. Wong, M.H.; Dobbins, T.A.; Tseung, J.; Tran, N.; Lee, C.S.; O'Brien, C.J.; Clark, J.; Rose, B.R. Oestrogen receptor β expression in pleomorphic adenomas of the parotid gland. *J. Clin. Pathol.* **2009**, *62*, 789–793. [CrossRef] [PubMed]
13. Kuiper, G.G.; Enmark, E.; Pelto-Huikko, M.; Nilsson, S.; Gustafsson, J.A. Cloning of a novel receptor expressed in rat prostate and ovary. *Proc. Natl. Acad. Sci. USA* **1996**, *93*, 5925–5930. [CrossRef] [PubMed]
14. Ohshiro, K.; Rayala, S.K.; Williams, M.D.; Kumar, R.; El-Naggar, A.K. Biological Role of Estrogen Receptor β in Salivary Gland Adenocarcinoma Cells. *Clin. Cancer Res.* **2006**, *12*, 5994–5999. [CrossRef] [PubMed]
15. Albanito, L.; Madeo, A.; Lappano, R.; Vivacqua, A.; Rago, V.; Carpino, A.; Oprea, T.I.; Prossnitz, E.R.; Musti, A.M.; Andò, S.; Maggiolini, M. G protein-coupled receptor 30 (GPR30) mediates gene expression changes and growth response to 17β-estradiol and selective GPR30 ligand G-1 in ovarian cancer cells. *Cancer Res.* **2007**, *67*, 1859–6610. [CrossRef] [PubMed]
16. Lappano, R.; Santolla, M.F.; Pupo, M.; Sinicropi, M.S.; Caruso, A.; Rosano, C.; Maggiolini, M. MIBE acts as antagonist ligand of both estrogen receptor α and GPER in breast cancer cells. *Breast Cancer Res.* **2012**, *14*, R12. [CrossRef] [PubMed]
17. Mau, M.; Mielenz, M.; Südekum, K.H.; Obukhov, A.G. Expression of GPR30 and GPR43 in oral tissues: Deriving new hypotheses on the role of diet in animal physiology and the development of oral cancers. *J. Anim. Physiol. Anim. Nutr.* **2011**, *95*, 280–285. [CrossRef] [PubMed]
18. Locati, L.D.; Perrone, F.; Losa, M.; Mela, M.; Casieri, P.; Orsenigo, M.; Cortelazzi, B.; Negri, T.; Tamborini, E.; Quattrone, P.; et al. Treatment relevant target immunophenotyping of 139 salivaryglandcarcinomas (SGCs). *Oral Oncol.* **2009**, *45*, 986–990. [CrossRef] [PubMed]
19. Kuhl, H. Pharmacology of estrogens and progestogens: Influence of different routes of administration. *Climacteric* **2005**, *1*, 3–63. [CrossRef] [PubMed]

20. Nasser, S.M.; Faquin, W.C.; Dayal, Y. Expression of androgen, estrogen, and progesterone receptors in salivary gland tumors. Frequent expression of androgen receptor in a subset of malignant salivary gland tumors. *Am. J. Clin. Pathol.* **2003**, *119*, 801–806. [CrossRef] [PubMed]

21. Barnes, L.; Rao, U.; Contis, L.; Krause, J.; Schwartz, A.; Scalamogna, P. Salivary duct carcinoma. Part II. Immunohistochemical evaluation of 13 cases for estrogen and progesterone receptors, cathepsin D, and C-ERBB-2 protein. *Oral. Surg. Oral. Med. Oral. Pathol.* **1994**, *78*, 74–80. [CrossRef]

22. Shick, P.C.; Riordan, G.P.; Foss, R.D. Estrogen and progesterone receptors in salivary gland adenoid cystic carcinoma. *Oral. Surg. Oral. Med. Oral. Pathol. Oral. Radiol. Endod.* **1995**, *80*, 440–444. [CrossRef]

23. Dori, S.; Trougouboff, P.; David, R.; Buchner, A. Immunohistochemical evaluation of estrogen and progesterone receptors in adenoid cystic carcinoma of salivary gland origin. *Oral Oncol.* **2000**, *36*, 450–453. [CrossRef]

24. Jeannon, J.P.; Soames, J.V.; Bell, H.; Wilson, J.A. Immunohistochemical detection of oestrogen and progesterone receptors in salivary tumours. *Clin. Otolaryngol. Allied Sci.* **1999**, *24*, 52–54. [CrossRef] [PubMed]

25. Pires, F.R.; da Cruz Perez, D.E.; de Almeida, O.P.; Kowalski, L.P. Estrogen receptor expression in salivary gland mucoepidermoid carcinoma and adenoid cystic carcinoma. *Pathol. Oncol. Res.* **2004**, *10*, 166–168. [CrossRef] [PubMed]

26. Williams, M.D.; Roberts, D.; Blumenschein, G.R., Jr.; Temam, S.; Kies, M.S.; Rosenthal, D.I.; Weber, R.S.; El-Naggar, A.K. Differential expression of hormonal and growth factor receptors in salivary duct carcinomas: Biologic significance and potential role in therapeutic stratification of patients. *Am. J. Surg. Pathol.* **2007**, *31*, 1645–1652. [CrossRef] [PubMed]

27. Sygut, D.; Bień, S.; Ziółkowska, M.; Sporny, S. Immunohistochemical expression of androgen receptor in salivary gland cancers. *Pol. J. Pathol.* **2008**, *59*, 205–210. [PubMed]

28. Moriki, T.; Ueta, S.; Takahashi, T.; Mitani, M.; Ichien, M. Salivary duct carcinoma: Cytologic characteristics and application of androgen receptor immunostaining for diagnosis. *Cancer* **2001**, *93*, 344–350. [CrossRef] [PubMed]

29. Nakajima, Y.; Kishimoto, T.; Nagai, Y.; Yamada, M.; Iida, Y.; Okamoto, Y.; Ishida, Y.; Nakatani, Y.; Ichinose, M. Expressions of androgen receptor and its co-regulators in carcinoma ex pleomorphic adenoma of salivary gland. *Pathology* **2009**, *41*, 634–639. [CrossRef] [PubMed]

30. Jaspers, H.C.; Verbist, B.M.; Schoffelen, R.; Mattijssen, V.; Slootweg, P.J.; van der Graaf Winette, T.A.; Carla, M.L.; van Herpen, C.M. Androgen receptor-positive salivary duct carcinoma: A disease entity with promising new treatment options. *J. Clin. Oncol.* **2011**, *29*, e473–e476. [CrossRef] [PubMed]

31. Välimaa, H.; Savolainen, S.; Soukka, T.; Silvoniemi, P.; Mäkelä, S.; Kujari, H.; Gustafsson, J.A.; Laine, M. Estrogen receptor-β is the predominant estrogen receptor subtype in human oral epithelium and salivary glands. *J. Endocrinol.* **2004**, *180*, 55–62. [CrossRef] [PubMed]

32. Locati, L.D.; Collini, P.; Imbimbo, M.; Barisella, M.; Testi, A.; Licitra, L.F.; Löning, T.; Tiemann, K.; Quattrone, P.; Bimbatti, E.; et al. Immunohistochemical and molecular profile of salivary gland cancer in children. *Pediatr. Blood Cancer* **2017**, *64*. [CrossRef] [PubMed]

33. Omoto, Y.; Iwase, H. Clinical significance of estrogen receptor β in breast and prostate cancer from biological aspects. *Cancer Sci.* **2015**, *106*, 337–343. [CrossRef] [PubMed]

34. Hopp, T.A.; Weiss, H.L.; Parra, I.S.; Cui, Y.; Osborne, C.K.; Fuqua, S.A. Low levels of estrogen receptor β protein predict resistance to tamoxifen therapy in breast cancer. *Clin. Cancer Res.* **2004**, *10*, 7490–7499. [CrossRef] [PubMed]

35. Iwase, H.; Zhang, Z.; Omoto, Y.; Sugiura, H.; Yamashita, H.; Toyama, T.; Iwata, H.; Kobayashi, S. Clinical significance of the expression of estrogen receptors alpha and β for endocrine therapy of breast cancer. *Cancer Chemother. Pharmacol.* **2003**, *52*, S34–S38. [CrossRef] [PubMed]

36. Murphy, L.C.; Peng, B.; Lewis, A.; Davie, J.R.; Leygue, E.; Kemp, A.; Ung, K.; Vendetti, M.; Shiu, S. Inducible upregulation of oestrogen receptor-β1 affects oestrogen and tamoxifen responsiveness in MCF7 human breast cancer cells. *J. Mol. Endocrinol.* **2005**, *34*, 553–566. [CrossRef] [PubMed]

37. Thomas, C.; Gustafsson, J.Å. The different roles of ER subtypes in cancer biology and therapy. *Nat Rev Cancer.* **2011**, *1*, 597–608. [CrossRef] [PubMed]

38. Shaaban, A.M.; Green, A.R.; Karthik, S.; Alizadeh, Y.; Hughes, T.A.; Harkins, L.; Ellis, I.O.; Robertson, J.F.; Paish, E.C.; Saunders, P.T.; Groome, N.P. Nuclear and cytoplasmic expression of ERβ1, ERβ2, and ERβ5 identifies distinct prognostic outcome for breast cancer patients. *Clin Cancer Res.* **2008**, *14*, 5228–5235. [CrossRef] [PubMed]

39. De Stefano, I.; Zannoni, G.F.; Prisco, M.G.; Fagotti, A.; Tortorella, L.; Vizzielli, G.; Mencaglia, L.; Scambia, G.; Gallo, D. Cytoplasmic expression of estrogen receptor β (ERβ) predicts poor clinical outcome in advanced serous ovarian cancer. *Gynecol Oncol.* **2011**, *122*, 573–579. [CrossRef] [PubMed]

40. Zannoni, G.F.; Prisco, M.G.; Vellone, V.G.; De Stefano, I.; Vizzielli, G.; Tortorella, L.; Fagotti, A.; Scambia, G.; Gallo, D. Cytoplasmic expression of oestrogen receptor β (ERβ) as a prognostic factor in vulvar squamous cell carcinoma in elderly women. *Histopathology.* **2011**, *59*, 909–917. [CrossRef] [PubMed]

41. Prossnitz, E.R.; Arterburn, J.B.; Sklar, L.A. GPR30: AG protein-coupled receptor for estrogen. *Mol. Cell. Endocrinol.* **2007**, *265*, 138–142. [CrossRef] [PubMed]

42. Gaudet, H.M.; Cheng, S.B.; Christensen, E.M.; Filardo, E.J. The G-protein coupled estrogen receptor, GPER: The inside and inside-out story. *Mol. Cell. Endocrinol.* **2015**, *3*, 207–219. [CrossRef] [PubMed]

43. Cheng, S.B.; Graeber, C.T.; Quinn, J.A.; Filardo, E.J. Retrograde transport of the transmembrane estrogen receptor, G-protein-coupled-receptor-30 (GPR30/GPER) from the plasma membrane towards the nucleus. *Steroids.* **2011**, *76*, 892–896. [CrossRef] [PubMed]

44. Samartzis, E.P.; Noske, A.; Meisel, A.; Varga, Z.; Fink, D.; Imesch, P. The G Protein-Coupled Estrogen Receptor (GPER) Is Expressed in Two Different Subcellular Localizations Reflecting Distinct Tumor Properties in Breast Cancer. *PLoS ONE* **2014**, *9*, e83296. [CrossRef] [PubMed]

45. Sjöström, M.; Hartman, L.; Grabau, D.; Fornander, T.; Malmström, P.; Nordenskjöld, B.; Sgroi, D.C.; Skoog, L.; Stål, O.; Leeb-Lundberg, L.M.; et al. Lack of G protein-coupled estrogen receptor (GPER) in the plasma membrane is associated with excellent long-term prognosis in breast cancer. *Breast Cancer Res. Treat.* **2014**, *145*, 61–71. [CrossRef] [PubMed]

46. Bologa, C.G.; Revankar, C.M.; Young, S.M.; Edwards, B.S.; Arterburn, J.B.; Kiselyov, A.S.; Parker, M.A.; Tkachenko, S.E.; Savchuck, N.P.; Sklar, L.A.; et al. Virtual and biomolecular screening converge on a selective agonist for GPR30. *Nat. Chem. Biol.* **2006**, *2*, 207–212. [CrossRef] [PubMed]

47. Dennis, M.K.; Burai, R.; Ramesh, C.; Petrie, W.K.; Alcon, S.N.; Nayak, T.K.; Bologa, C.G.; Leitao, A.; Brailoiu, E.; Deliu, E.; et al. In vivo effects of a GPR30 antagonist. *Nat. Chem. Biol.* **2009**, *5*, 421–427. [CrossRef] [PubMed]

48. Dennis, M.K.; Field, A.S.; Burai, R.; Ramesh, C.; Petrie, W.K.; Bologa, C.G.; Oprea, T.I.; Yamaguchi, Y.; Hayashi, S.; Sklar, L.A.; et al. Identification of a GPER/GPR30 Antagonist with Improved Estrogen Receptor Counter selectivity. *J. Steroid Biochem. Mol. Biol.* **2011**, *127*, 358–366. [CrossRef] [PubMed]

49. Petrie, W.K.; Dennis, M.K.; Hu, C.; Dai, D.; Arterburn, J.B.; Smith, H.O.; Hathaway, H.J.; Prossnitz, E.R. G Protein-Coupled Estrogen Receptor-Selective Ligands Modulate Endometrial Tumor Growth. *Obstet. Gynecol. Int.* **2013**, *2013*, 472720. [CrossRef] [PubMed]

 International Journal of
Molecular Sciences

Article

Expression of Phospho-ELK1 and Its Prognostic Significance in Urothelial Carcinoma of the Upper Urinary Tract

Satoshi Inoue [1,2,3], Hiroki Ide [3], Kazutoshi Fujita [4], Taichi Mizushima [1,2,3], Guiyang Jiang [1,2], Takashi Kawahara [1,3], Seiji Yamaguchi [5], Hiroaki Fushimi [6], Norio Nonomura [4] and Hiroshi Miyamoto [1,2,3,7,*]

[1] Department of Pathology & Laboratory Medicine, University of Rochester Medical Center, Rochester, NY 14642, USA; inosts411@gmail.com (S.I.); mizu123shima@gmail.com (T.M.); guiyang_jiang@urmc.rochester.edu (G.J.); takashi_tk2001@yahoo.co.jp (T.K.)
[2] James P. Wilmot Cancer Center, University of Rochester Medical Center, Rochester, NY 14642, USA
[3] Department of Pathology and James Buchanan Brady Urological Institute, Johns Hopkins University School of Medicine, Baltimore, MD 21287, USA; h-ide@fc4.so-net.ne.jp
[4] Department of Urology, Osaka University Graduate School of Medicine, Suita 565-0871, Japan; fujita@uro.med.osaka-u.ac.jp (K.F.); nono@uro.med.osaka-u.ac.jp (N.N.)
[5] Department of Urology, Osaka General Medical Center, Osaka 558-8558, Japan; yamabu1956@gmail.com
[6] Department of Pathology, Osaka General Medical Center, Osaka 558-8558, Japan; hiroaki-fushimi@gh.opho.jp
[7] Department of Urology, University of Rochester Medical Center, Rochester, NY 14642, USA
* Correspondence: hiroshi_miyamoto@urmc.rochester.edu; Tel.: +1-585-275-8748

Received: 20 February 2018; Accepted: 6 March 2018; Published: 8 March 2018

Abstract: Using preclinical models, we have recently found that ELK1, a transcriptional factor that activates downstream targets, including *c-fos* proto-oncogene, induces bladder cancer outgrowth. Here, we immunohistochemically determined the expression status of phospho-ELK1, an activated form of ELK1, in upper urinary tract urothelial carcinoma (UUTUC). Overall, phospho-ELK1 was positive in 47 (47.5%; 37 weak (1+) and 10 moderate (2+)) of 99 UUTUCs, which was significantly ($P = 0.002$) higher than in benign urothelium (21 (25.3%) of 83; 17 1+ and 4 2+) and was also associated with androgen receptor expression ($P = 0.001$). Thirteen (35.1%) of 37 non-muscle-invasive versus 34 (54.8%) of 62 muscle-invasive UUTUCs ($P = 0.065$) were immunoreactive for phospho-ELK1. Lymphovascular invasion was significantly ($P = 0.014$) more often seen in phospho-ELK1(2+) tumors (80.0%) than in phospho-ELK1(0/1+) tumors (36.0%). There were no statistically significant associations between phospho-ELK1 expression and tumor grade, presence of concurrent carcinoma in situ or hydronephrosis, or pN status. Kaplan-Meier and log-rank tests revealed that patients with phospho-ELK1(2+) tumor had marginally and significantly higher risks of disease progression ($P = 0.055$) and cancer-specific mortality ($P = 0.008$), respectively, compared to those with phospho-ELK1(0/1+) tumor. The current results thus support our previous observations in bladder cancer and further suggest that phospho-ELK1 overexpression serves as a predictor of poor prognosis in patients with UUTUC.

Keywords: ELK1; immunohistochemistry; upper urinary tract urothelial carcinoma; prognosis; androgen receptor

---

1. Introduction

Upper urinary tract urothelial carcinoma (UUTUC) is a relatively rare disease accounting for only 5–10% of all urothelial carcinomas, whereas urothelial carcinoma of the urinary bladder is a common malignancy, especially in males [1,2]. Due to its preponderance, clinical evidence for bladder cancer has often been applied to decision-making on UUTUC. Indeed, only a few major urological or oncologic associations (e.g., European Association of Urology, Japanese Urological Association) have published a guideline for UUTUC separate from that for bladder cancer [1,2]. More strikingly, there are no prognostic markers for UUTUC available for clinical practice, while alterations in some molecular or genetic factors, which are associated with bladder cancer and serve as its prognosticators, are observed in UUTUC [3–5].

ELK1, as a transcription factor, is phosphorylated through activating the mitogen-activated protein kinase (MAPK)/extracellular signal-regulated kinase (ERK) pathways and translocates to the nucleus, leading to the regulation of downstream targets, including a proto-oncogene *c-fos* [6,7], as well as matrix metalloproteinases (MMPs) [8,9] that contribute to tumor cell invasion. Using in vitro and in vivo models for bladder cancer, we have recently found that ELK1 activation correlates with the induction of cell proliferation, migration, and invasion, as well as resistance to cisplatin cytotoxicity [10,11]. Meanwhile, emerging preclinical evidence has indicated a critical role of androgen-mediated androgen receptor (AR) signaling in the development and progression of urothelial cancer [12]. Interestingly, ELK1 appeared to require a functional AR for inducing cell proliferation [10,11]. Indeed, in prostate cancer cells, AR has been shown to function as a co-activator of ELK1 [13]. In surgical specimens, we also demonstrated that ELK1 or phospho-ELK1 (p-ELK1) expression was up-regulated in bladder cancer, compared with non-neoplastic urothelium, and that positivity of p-ELK1, but not ELK1, was associated with the risk of recurrence of non-muscle-invasive tumors (hazard ratio (HR) = 2.829; $P = 0.056$) or cancer-specific mortality in patients with muscle-invasive tumor (HR = 2.693; $P = 0.021$) in a multivariate setting [11].

Thus, ELK1 has been suggested to not only promote urothelial cancer progression, but also function as an important prognosticator for bladder cancer. By contrast, the status of ELK1 expression in UUTUC and its prognostic significance remained uncertain. The aim of this study was to examine the association between p-ELK1 expression and clinicopathological features of UUTUC.

2. Results

2.1. Immunoreactivity in Benign and Tumor Tissues

Using immunohistochemistry, we investigated the expression of an activated form of ELK1, p-ELK1, in a tissue microarray (TMA) consisting of 99 UUTUC specimens, as well as 83 corresponding normal-appearing urothelial tissue samples. Positive signals for p-ELK1 were detected predominantly in the nuclei of non-neoplastic (Figure 1a) and neoplastic (Figure 1b) epithelial cells. The status of p-ELK1 expression in benign versus tumor tissues is summarized in Table 1. p-ELK1 was positive in 21 (25.3%) of 83 benign urothelial tissues (17 (20.5%) weak (1+) and 4 (4.8%) moderate (2+)) and 47 (47.5%) of 99 urothelial neoplasms (37 (37.4%) 1+ and 10 (10.1%) 2+). Thus, the rate of p-ELK1 positivity was significantly higher in tumors than in benign tissues ($P = 0.002$).

Table 1. p-ELK1 expression in non-neoplastic urothelium versus urothelial neoplasm tissue specimens.

Tissue	*n*	p-ELK1 Expression				*P* Value	
		0 (%)	1+ (%)	2+ (%)	3+ (%)	0 vs. 1+/2+/3+	0/1+ vs. 2+/3+
Normal	83	62 (74.7)	17 (20.5)	4 (4.8)	0 (0)	0.002	0.265
Tumor	99	52 (52.5)	37 (37.4)	10 (10.1)	0 (0)		

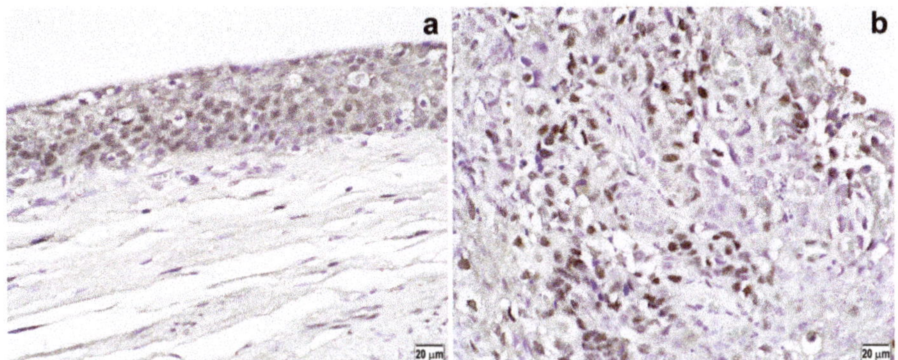

Figure 1. Immunohistochemistry of p-ELK1 in normal urothelial tissue (**a**) and urothelial tumor (**b**). A semi-quantitative analysis of p-ELK1 expression is performed by employing a combination of staining intensity (i.e., weak (**a**), strong (**b**)) and distribution (i.e., percentage of immunoreactive cells). Original magnification: 400×.

2.2. Immunoreactivity and Clinicopathological Features

The status of p-ELK1 expression in UUTUCs according to clinicopathological features is shown in Table 2. The p-ELK1 expression levels tended to be elevated in muscle-invasive tumors, compared with non-muscle-invasive tumors, but they were not statistically different between low-grade and high-grade carcinomas. Lymphovascular invasion was significantly ($P = 0.014$) more often seen in p-ELK1(2+) tumors (8 of 10 (80.0%)) than in p-ELK1(0/1+) tumors (32 of 89 (36.0%)). However, other features, including patient age or gender, tumor laterality, presence of concurrent carcinoma in situ or hydronephrosis, and lymph node involvement, were not significantly associated with p-ELK1 expression. As for tumor site, moderate (2+) p-ELK1 expression was marginally more often ($P = 0.096$) seen in ureteral tumors, compared with renal pelvic tumors. The rates of p-ELK1 positivity in the renal pelvic tumors, ureteral tumors, and bladder tumors were 40.0% (18 of 45), 56.0% (28 of 50), and 65.9% (85 of 129; shown in our previous study [11]), respectively (renal pelvis vs. bladder: $P = 0.003$; ureter vs. bladder: $P = 0.231$).

Table 2. Correlations between p-ELK1 expression and clinicopathological profile of the patients.

Parameter	n	p-ELK1 Expression			p Value	
		0 (%)	1+ (%)	2+ (%)	0 vs. 1+/2+	0/1+ vs. 2+
Age (mean ± SD; years)	99	70.0 ± 9.5	71.9 ± 7.3	68.5 ± 10.9	0.199	0.659
Gender					0.849	0.736
Male	60	28 (46.7)	25 (41.7)	7 (11.7%)		
Female	39	24 (61.5)	12 (30.8)	3 (7.7)		
Laterality					0.548	0.323
Right	43	21 (48.8)	16 (37.2)	6 (14.0)		
Left	56	31 (55.4)	21 (37.5)	4 (7.1)		
Tumor site					0.151 [a]	0.096 [a]
Renal pelvis	45	27 (60.0)	16 (35.6)	2 (4.4)		
Ureter	50	22 (44.0)	20 (40.0)	8 (16.0)		
Both	4	3 (75.0)	1 (25.0)	0 (0)		
Tumor grade					0.273	1.000
Low-grade	15	10 (66.7)	4 (26.7)	1 (6.7)		
High-grade	84	42 (50.0)	33 (39.3)	9 (10.7)		
Pathologic stage					0.065 [b]	0.085 [b]
pTa	19	13 (68.4)	5 (26.3)	1 (5.3)		
pT1	18	11 (61.1)	7 (38.9)	0 (0)		
NMI (pTa + pT1)	37	24 (64.9)	12 (32.4)	1 (2.7)		
pT2	8	1 (12.5)	6 (75.0)	1 (12.5)		
pT3	48	26 (54.2)	16 (33.3)	6 (12.5)		
pT4	6	1 (16.7)	3 (50.0)	2 (33.3)		
MI (pT2 + pT3 + pT4)	62	28 (45.2)	25 (40.3)	9 (14.5)		

Table 2. *Cont.*

Parameter	n	p-ELK1 Expression			p Value	
		0 (%)	1+ (%)	2+ (%)	0 vs. 1+/2+	0/1+ vs. 2+
Concurrent CIS					0.768	0.616
No	86	46 (53.5)	32 (37.2)	8 (9.3)		
Yes	13	6 (46.2)	5 (38.5)	2 (15.4)		
Hydronephrosis					0.445 [c]	1.000 [c]
No	61	33 (54.1)	25 (41.0)	3 (4.9)		
Yes	20	13 (65.0)	6 (30.0)	1 (5.0)		
Unknown	18	6 (33.3)	6 (33.3)	6 (33.3)		
Lymphovascular invasion					0.227	0.014
No	59	34 (57.6)	23 (39.0)	2 (3.4)		
Yes	40	18 (45.0)	14 (35.0)	8 (20.0)		
Lymph node involvement					0.357 [d]	0.109 [d]
pN0	84	41 (48.8)	36 (42.9)	7 (8.3)		
pN1-3	12	8 (66.7)	1 (8.3)	3 (25.0)		
pNx	3	3 (100)	0 (0)	0 (0)		

NMI = non-muscle-invasive; MI = muscle-invasive; CIS = carcinoma in situ. [a] Renal pelvis vs. ureter; [b] NMI vs. MI; [c] No vs. Yes; [d] pN0 vs. pN1-3.

We then analyzed the relationship between the positivity of p-ELK1 and steroid hormone receptors including AR, estrogen receptor (ER)-α, ERβ, glucocorticoid receptor (GR), and progesterone receptor (PR). Using the same cohort of 99 patients, we reported that AR/ERα/ERβ/GR/PR were positive in 20 (20.2%)/18 (18.2%)/62 (62.6%)/62 (62.6%)/16 (16.2%) UUTUCs, respectively [14]. There was a tendency to show a weak positive correlation (i.e., correlation coefficient (CC) = 0.2–0.4) between p-ELK1 and AR positivity, especially in male tumors (CC = 0.247; P = 0.058) (Table 3). Thus, of 52 p-ELK1-negative vs. 47 p-ELK1-positive tumors, 4 (7.7%) vs. 16 (34.0%) were positive for AR (P = 0.001). Similarly, of 26 p-ELK1-negative vs. 34 P-ELK1-positive male tumors, 2 (7.7%) vs. 16 (41.2%) were positive for AR (P = 0.007). No significant correlations between p-ELK1 and ERα, ERβ, GR, or PR were seen in all 99 tumors, 60 male tumors, and 39 female tumors.

Table 3. Correlations between p-ELK1 and AR/ERα/ERβ/GR/PR expression.

Patients	n	AR		ERα		ERβ		GR		PR	
		CC	P	CC	P	CC	P	CC	P	CC	P
All cases	99	0.171	0.091	0.076	0.454	0.103	0.312	0.176	0.081	0.054	0.594
Male	60	0.247	0.058	0.175	0.181	0.082	0.535	0.096	0.466	0.055	0.678
Female	39	−0.105	0.525	−0.048	0.770	0.199	0.224	0.262	0.107	0.137	0.407

2.3. Immunoreactivity and Prognostic Significance

Next, we investigated possible associations between p-ELK1 expression and patient outcomes. To more accurately assess the role of p-ELK1 expression in disease progression, those with M1 disease (n = 4) at the time of nephroureterectomy were excluded from the analyses. Kaplan-Meier and log-rank tests revealed no significant associations between p-ELK1 levels and tumor recurrence in the bladder (0 vs. 1+/2+, P = 0.458; 0/1+ vs. 2+, P = 0.806). By contrast, moderate p-ELK1 expression was marginally or significantly associated with lower progression-free survival (PFS) (0/1+ vs. 2+, P = 0.055; Figure 2a,b), overall survival (OS) (0/1+ vs. 2+, P = 0.020; figure not shown), and cancer-specific survival (CSS) (0/1+ vs. 2+, P = 0.008; Figure 2c,d) rates. Significant differences in the prognosis were still seen in 81 cases of high-grade tumors (PFS: P = 0.042; OS: P = 0.022; CSS: P = 0.013), but not in 58 cases of muscle-invasive tumors (PFS: P = 0.411; OS: P = 0.163; CSS: P = 0.135).

To determine if p-ELK1 expression status was an independent prognosticator for UUTUC, we then performed multivariate analysis, using the Cox model, for the factors showing P < 0.1 in univariate analysis (Table 4). In 95 patients without M1 disease, pT stage and lymphovascular invasion were associated with PFS and/or CSS. However, no significant associations between p-ELK1 expression versus PFS or CSS were found.

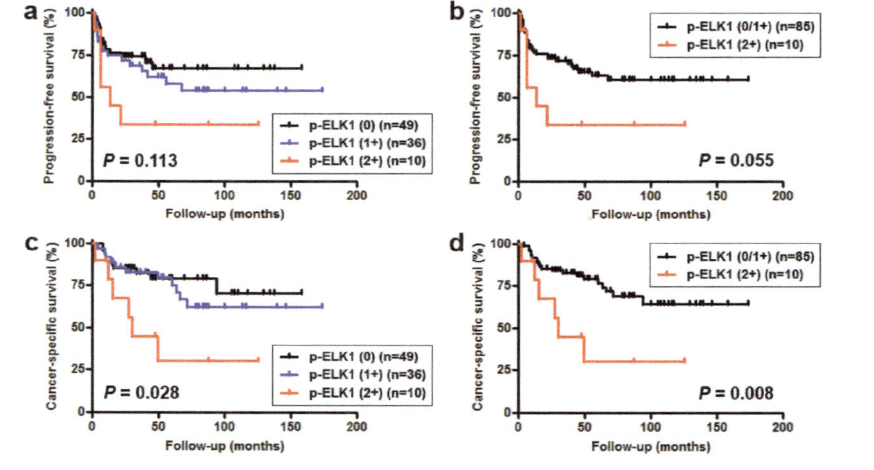

Figure 2. Kaplan-Meier curves for PFS (**a**,**b**) or CSS (**c**,**d**) in 95 patients without metastatic disease, according to the status of p-ELK1 expression.

Table 4. Univariate and multivariate analysis of PFS and CSS in 95 patients with UUTUC.

| Parameter | Progression-Free Survival | | | | | | Cancer-Specific Survival | | | | | |
| | Univariate | | | Multivariate | | | Univariate | | | Multivariate | | |
	HR	95% CI	*P*	HR	95% CI	*P*	HR	95% CI	*P*	HR	95% CI	*P*
Tumor grade	3.858	0.923–16.123	0.064	3.304	0.715–12.877	0.132	6.411	0.868–47.372	0.036	4.953	0.661–37.086	0.119
pT stage [a]	10.975	3.848–31.306	<0.001	7.750	2.575–23.329	<0.001	17.213	4.055–73.070	<0.001	10.118	2.241–45.680	0.003
LVI	5.701	2.775–11.711	<0.001	2.483	1.125–5.481	0.024	6.712	2.827–15.934	<0.001	2.350	0.888–6.222	0.085
pN stage	4.232	1.738–10.308	0.001	2.494	0.891–6.981	0.082	4.379	1.762–10.884	0.001	1.603	0.605–4.244	0.343
p-ELK1 [b]	2.291	0.948–5.540	0.066	0.666	0.244–1.820	0.428	3.179	1.279–7.901	0.013	1.131	0.431–2.964	0.802

LVI = lymphovascular invasion; HR = hazard ratio; CI = confidence interval. [a] pTa-2 vs. pT3-4; [b] 0/1+ vs. 2+.

3. Discussion

The functional role of ELK1, an upstream regulator of the *c-fos* oncogene, in the development and progression of UUTUC remains poorly understood. In the present study, we immunohistochemically determined the expression status of p-ELK1 in UUTUC specimens and its prognostic significance. We first compared the levels of p-ELK1 expression in tumors versus adjacent normal tissues in the upper urinary tract. In accordance with our observations in bladder specimens [11], p-ELK1 expression was significantly up-regulated in tumors, compared with the non-neoplastic urothelium. These results may suggest that ELK1 activation contributes to urothelial tumorigenesis at both the upper and lower urinary tracts. Indeed, we recently found, using an in vitro system, that ELK1 signals were associated with the induction of neoplastic transformation of urothelial cells (Inoue et al., unpublished data).

ELK1 has been implicated in the regulation of cell proliferation, cell cycle control, apoptosis, and cell migration/invasion via, for instance, activation of MAPK/ERK signaling [6,7,15,16]. It also modulates the expression of MMPs [8,9]. Here, we further demonstrated that p-ELK1 overexpression was marginally and significantly associated with muscle invasion and lymphovascular invasion, respectively, in UUTUC. Univariate analysis revealed that p-ELK1 overexpression was also marginally and significantly associated with disease progression and cancer-specific mortality, respectively, in patients with UUTUC. The current findings not only are consistent with those in bladder specimens, indicating the prognostic values of p-ELK1 expression in patients with muscle-invasive tumor [11], but also support our observations in preclinical models suggesting that ELK1 promotes the proliferation, migration, and invasion of bladder cancer cells and activates MMP-2 and MMP-9 [10,11]. Thus, ELK1 activity is suggested to predict the prognosis of UUTUC. However, multivariate analysis did not show statistical significance for p-ELK1 overexpression. In addition, although p-ELK1 positivity in non-muscle-invasive bladder tumors was shown to predict the risk of their recurrence [11], we failed to show an association between p-ELK1 overexpression in UUTUCs and their recurrence in the bladder.

The functional interactions between ELK1 and AR signaling pathways have been documented in prostate cancer cells [13]. We also previously demonstrated activation of ELK1 by androgen-mediated AR signals in bladder cancer cells, as well as a significant association between the expression levels of p-ELK1 and AR in bladder tumor tissue specimens [11]. Moreover, ELK1 inactivation resulted in strong inhibition of the growth of bladder (and prostate) cancer cells only in the presence of an activated AR [10,11,17]. We here showed a marginal association between p-ELK1 and AR expression in UUTUC samples. These findings suggest the involvement of AR signaling in the induction of urothelial cancer progression by ELK1. No significant associations of p-ELK1 expression with that of other steroid hormone receptors, including ERα, ERβ, GR, and PR.

The levels of p-ELK1 expression were higher in ureteral tumors than in renal pelvic tumors, as well as in bladder tumors [11], than in ureteral tumors. Using immunohistochemistry in the same sets of UUTUC and bladder cancer TMAs, we have assessed the expression of various proteins. Interestingly, renal pelvic tumors, compared with ureteral tumors, exhibited lower positive rates of five (out of seven) transcription factors, including AR (11.1% vs. 28.0%, $P = 0.070$) [14], ERβ (51.1% vs. 72.0%, $P = 0.056$) [14], GR (57.8% vs. 68.0%, $P > 0.1$) [14], GATA3 (35.6% vs. 66.0%, $P = 0.004$) [18], and ZKSCAN3 (26.7% vs. 54.0%, $P = 0.012$) [19]. The expression of all of these transcription factors (except ERβ), in addition to p-ELK1, was further up-regulated in bladder tumors, compared with ureteral tumors [20–23]. The underlying reasons for these findings in the expression of p-ELK1 and other transcription factors in renal pelvic tumors vs. ureteral tumors vs. bladder tumors remain undefined. Of note, the expression patterns of these transcription factors are not well correlated with their functional roles (e.g., tumor suppressive vs. oncogenic) since it has been documented that some promote and others inhibit urothelial cancer outgrowth. However, as we previously suggested [14,24], differences in the anatomic location of renal pelvic/ureteral/bladder tumors and the thickness of the specimens around the tumors might have affected the immunoreactivity for p-ELK1, owing to, for instance, those in the time to complete tissue fixation. Another possibility includes a higher proportion of muscle-invasive disease, where p-ELK1 expression is more likely stronger, in ureteral tumors (33 of 50 (66.0%)) than in renal pelvic tumors (25 of 45 (55.6%)). Meanwhile, the rates of p-ELK1 positivity were similar between benign portions of renal pelvic (11 of 38 (28.9%)) versus ureteral (10 of 41 (24.4%)) urothelium. This may still be because urothelial tissues of both the renal pelvis and ureter are located on the surface of nephroureterectomy specimens when they are opened for gross examination and fixation.

4. Materials and Methods

4.1. Patients and Tissue Samples

Upon the approval by the institutional review board (IRB #25-2014 at Osaka General Medical Center, Osaka, Japan; Date: 19 June 2013), UUTUC TMA was constructed, as we described previously [25], consisting of dominant tumors and paired normal-appearing urothelial tissues from patients undergoing radical nephroureterectomy. Clinicopathological data of these 99 patients were described previously [14,25] (also see Table 2). There were four cases with metastatic disease where nephroureterectomy was performed mainly for bleeding control. None of the patients had received therapy with anti-cancer drugs or radiation preoperatively.

4.2. Immunohistochemistry

Immunohistochemical staining was carried out on the 5 μm sections from the UUTUC TMA, using a primary antibody to p-ELK1 (Ser^{383}; sc-8406; Santa Cruz Biotechnology), as we described previously [10,11]. Two pathologists (Guiyang Jiang and Hiroshi Miyamoto), who were blinded to patient identity, independently scored only nuclear staining, using the German Immunoreactive Score (0–12), calculated by multiplying the percentage of immunoreactive cells (0% = 0; 1–10% = 1; 11–50% = 2; 51–80% = 3; 81–100% = 4) by staining intensity (negative = 0; weak = 1; moderate = 2; strong = 3). The scores of 0–1, 2–4, 6–8, and 9–12 were then considered negative (0), weakly positive (1+), moderately positive (2+), and strongly positive (3+), respectively. Cases with discrepancies were re-reviewed simultaneously by the two pathologists until a consensus was reached.

4.3. Statistical Analyses

The Fisher's exact test and Mann–Whitney U test were used to assess the statistical significance for categorized variables and those with ordered distribution, respectively. Correlations between variables were determined by the Spearman's correlation. The rates of recurrence-free survival, PFS, and CSS were calculated by the Kaplan-Meier method, and differences were analyzed by the log-rank test. The Cox proportional hazards model was used for multivariate analysis of prognosticators. P values less than 0.05 were considered statistically significant.

5. Conclusions

We showed a significant increase in the expression of p-ELK1 in UUTUC, compared with normal-appearing urothelium from each case, implying the involvement of ELK1 signals in the outgrowth of UUTUC. The current results also support our in vitro and in vivo findings in bladder cancer and further suggest that p-ELK1 overexpression serves as a predictor of poor prognosis in patients with UUTUC. Further studies with larger patient cohort are required to validate our observations.

Author Contributions: Satoshi Inoue and Hiroshi Miyamoto conceived and designed the experiments; Satoshi Inoue, Hiroki Ide, Taichi Mizushima, and Guiyang Jiang performed the experiments; Satoshi Inoue, Kazutoshi Fujita, and Taichi Mizushima analyzed the data; Taichi Mizushima, Takashi Kawahara, Seiji Yamaguchi, Hiroaki Fushimi, and Norio Nonomura contributed reagents/materials/analysis tools; Satoshi Inoue drafted the manuscript and Hiroshi Miyamoto edited it. All authors reviewed the manuscript.

Conflicts of Interest: The authors declare no conflict of interest.

References

1. Oya, M.; Kikuchi, E.; Committee for Establishment of Clinical Practice Guideline for Management of Upper Tract Urothelial Carcinoma and Japanese Urological Association. Evidenced-based clinical practice guideline for upper tract urothelial carcinoma (summary–Japanese Urological Association, 2014 edition). *Int. J. Urol.* **2015**, *22*, 3–13. [PubMed]
2. Rouprêt, M.; Babjuk, M.; Compérat, E.; Zigeuner, R.; Sylvester, R.J.; Burger, M.; Cowan, N.C.; Böhle, A.; Van Rhijn, B.W.; Kaasinen, E.; et al. European Association of Urology guidelines on upper urinary tract urothelial cell carcinoma: 2015 Update. *Eur. Urol.* **2015**, *68*, 868–879. [CrossRef] [PubMed]
3. Chromecki, T.F.; Bensalah, K.; Remzi, M.; Verhoest, G.; Cha, E.K.; Scherr, D.S.; Novara, G.; Karakiewicz, P.I.; Shariat, S.F. Prognostic factors for upper urinary tract urothelial carcinoma. *Nat. Rev. Urol.* **2011**, *8*, 440–447. [CrossRef] [PubMed]
4. Lughezzani, G.; Burger, M.; Margulis, V.; Matin, S.F.; Novara, G.; Roupret, M.; Shariat, S.F.; Wood, C.G.; Zigeuner, R. Prognostic factors in upper urinary tract urothelial carcinomas: A comprehensive review of the current literature. *Eur. Urol.* **2012**, *62*, 100–114. [CrossRef] [PubMed]
5. Krabbe, L.M.; Heitplatz, B.; Preuss, S.; Hutchinson, R.C.; Woldu, S.L.; Singla, N.; Boegemann, M.; Wood, C.G.; Karam, J.A.; Weizer, A.Z.; et al. Prognostic value of PD-1 and PD-L1 expression in patients with high grade upper tract urothelial carcinoma. *J. Urol.* **2017**, *198*, 1253–1262. [CrossRef] [PubMed]
6. Hipskind, R.A.; Rao, V.N.; Mueller, C.G.; Reddy, E.S.; Nordheim, A. Ets-related protein Elk-1 is homologous to the c-fos regulatory factor p62TCF. *Nature* **1991**, *354*, 531–534. [CrossRef] [PubMed]
7. Vanhoutte, P.; Bamier, J.V.; Guibert, B.; Pagès, C.; Besson, M.J.; Hipskind, R.A.; Caboche, J. Glutamate induces phosphorylation of Elk-1 and CREB, along with c-fos activation, via an extracellular signal-regulated kinase-dependent pathway in brain slices. *Mol. Cell. Biol.* **1999**, *19*, 136–146. [CrossRef] [PubMed]
8. Choi, B.D.; Jeong, S.J.; Wang, G.; Park, J.J.; Lim, D.S.; Kim, B.H.; Cho, Y.I.; Kim, C.S.; Jeong, M.J. Secretory leukocyte protease inhibitor is associated with MMP-2 and MMP-9 to promote migration and invasion in SNU638 gastric cancer cells. *Int. J. Mol. Med.* **2011**, *28*, 527–534. [PubMed]
9. Ahmad, N.; Wang, W.; Nair, R.; Kapila, S. Relaxin induces matrix-metalloproteinases-9 and -13 via RXFP1: Induction of MMP-9 involves the PI3K, ERK, Akt and PKC-ζ pathways. *Mol. Cell. Endocrinol.* **2012**, *363*, 46–61. [CrossRef] [PubMed]
10. Kawahara, T.; Ide, H.; Kashiwagi, E.; Patterson, J.D.; Inoue, S.; Shareef, H.K.; Aljarah, A.K.; Zheng, Y.; Baras, A.S.; Miyamoto, H. Silodosin inhibits the growth of bladder cancer cells and enhances the cytotoxic activity of cisplatin via ELK1 inactivation. *Am. J. Cancer Res.* **2015**, *5*, 2959–2968. [PubMed]
11. Kawahara, T.; Shareef, H.K.; Aljarah, A.K.; Ide, H.; Li, Y.; Kashiwagi, E.; Netto, G.J.; Zheng, Y.; Miyamoto, H. ELK1 is up-regulated by androgen in bladder cancer cells and promotes tumor progression. *Oncotarget* **2015**, *6*, 29860–29876. [CrossRef] [PubMed]
12. Inoue, S.; Mizushima, T.; Miyamoto, H. Role of the androgen receptor in urothelial cancer. *Mol. Cell. Endocrinol.* **2017**. [CrossRef] [PubMed]
13. Patki, M.; Chari, V.; Sivakumaran, S.; Gonit, M.; Trumbly, R.; Ratnam, M. The ETS domain transcription factor ELK1 directs a critical component of growth signaling by the androgen receptor in prostate cancer cells. *J. Biol. Chem.* **2013**, *288*, 11047–11065. [CrossRef] [PubMed]
14. Kashiwagi, E.; Fujita, K.; Yamaguchi, S.; Fushimi, H.; Ide, H.; Inoue, S.; Mizushima, T.; Reis, L.O.; Sharma, R.; Netto, G.J.; et al. Expression of steroid hormone receptors and its prognostic significance in urothelial carcinoma of the upper urinary tract. *Cancer Biol. Ther.* **2016**, *17*, 1188–1196. [CrossRef] [PubMed]
15. Shao, N.; Chai, Y.; Cui, J.Q.; Wang, N.; Aysola, K.; Reddy, E.S.; Rao, V.N. Induction of apoptosis by Elk-1 and ΔElk-1 proteins. *Oncogene* **1998**, *17*, 527–532. [CrossRef] [PubMed]
16. Mut, M.; Lule, S.; Demir, O.; Kumaz, I.A.; Vural, I. Both mitogen-activated protein kinase (MAPK)/extracellular-signal-regulated kinases (ERK) 1/2 and phosphatidylinositide-3-OH kinase (PI3K)/Akt pathways regulate activation of E-twenty-six (ETS)-like transcription factor 1 (Elk-1) in U138 glioblastoma cells. *Int. J. Biochem. Cell Biol.* **2012**, *44*, 302–310. [CrossRef] [PubMed]
17. Kawahara, T.; Aljarah, A.K.; Shareef, H.K.; Inoue, S.; Ide, H.; Patterson, J.D.; Kashiwagi, E.; Han, B.; Li, Y.; Zheng, Y.; et al. Silodosin inhibits prostate cancer cell growth via ELK1 inactivation and enhances the cytotoxic activity of gemcitabine. *Prostate* **2016**, *76*, 744–756. [CrossRef] [PubMed]

18. Inoue, S.; Mizushima, T.; Fujita, K.; Meliti, A.; Ide, H.; Yamaguchi, S.; Fushimi, H.; Netto, G.J.; Nonomura, N.; Miyamoto, H. GATA3 immunohistochemistry in urothelial carcinoma of the upper urinary tract as a urothelial marker as well as a prognosticator. *Hum. Pathol.* **2017**, *64*, 83–90. [CrossRef] [PubMed]

19. Jalalizadeh, M.; Inoue, S.; Fujita, K.; Ide, H.; Mizushima, T.; Yamaguchi, S.; Fushimi, H.; Nonomura, N.; Miyamoto, H. ZKSCAN3 expression in urothelial carcinoma of the upper urinary tract and its impact on patient outcomes. *Integr. Cancer Sci. Ther.* **2017**, *4*, 1000241. [CrossRef]

20. Miyamoto, H.; Izumi, K.; Yao, J.L.; Li, Y.; Yang, Q.; McMahon, L.A.; Gonzalez-Roibon, N.; Hicks, D.G.; Tacha, D.; Netto, G.J. GATA binding protein 3 is down-regulated in bladder cancer yet strong expression is an independent predictor of poor prognosis in invasive tumor. *Hum. Pathol.* **2012**, *43*, 2033–2040. [CrossRef] [PubMed]

21. Miyamoto, H.; Yao, J.L.; Chaux, A.; Zheng, Y.; Hsu, I.; Izumi, K.; Chang, C.; Messing, E.M.; Netto, G.J.; Yeh, S. Expression of androgen and oestrogen receptors and its prognostic significance in urothelial neoplasm of the urinary bladder. *BJU Int.* **2012**, *109*, 1716–1726. [CrossRef] [PubMed]

22. Ishiguro, H.; Kawahara, T.; Zheng, Y.; Netto, G.J.; Miyamoto, H. Reduced glucocorticoid receptor expression predicts bladder tumor recurrence and progression. *Am. J. Clin. Pathol.* **2014**, *142*, 157–164. [CrossRef] [PubMed]

23. Kawahara, T.; Inoue, S.; Ide, H.; Kashiwagi, E.; Ohtake, S.; Mizushima, T.; Li, P.; Li, Y.; Zheng, Y.; Uemura, H.; et al. ZKSCAN3 promotes bladder cancer cell proliferation, migration, and invasion. *Oncotarget* **2016**, *7*, 53599–53610. [CrossRef] [PubMed]

24. Kawahara, T.; Inoue, S.; Fujita, K.; Mizushima, T.; Ide, H.; Yamaguchi, S.; Fushimi, H.; Nonomura, N.; Miyamoto, H. NFATc1 expression as a prognosticator in urothelial carcinoma of the upper urinary tract. *Transl. Oncol.* **2017**, *10*, 318–323. [CrossRef] [PubMed]

25. Munari, E.; Fujita, K.; Faraj, S.; Chaux, A.; Gonzalez-Roibon, N.; Hicks, J.; Meeker, A.; Nonomura, N.; Netto, G.J. Dysregulation of mammalian target of rapamycin pathway in upper tract urothelial carcinoma. *Hum. Pathol.* **2013**, *44*, 2668–2676. [CrossRef] [PubMed]

International Journal of
Molecular Sciences

Interaction of ERα and NRF2 Impacts Survival in Ovarian Cancer Patients

Bastian Czogalla [1,†], Maja Kahaly [1,†], Doris Mayr [2], Elisa Schmoeckel [2], Beate Niesler [3], Thomas Kolben [1], Alexander Burges [1], Sven Mahner [1], Udo Jeschke [1,*]

[1] Department of Obstetrics and Gynecology, University Hospital, LMU Munich, 81377 Munich, Germany; bastian.czogalla@med.uni-muenchen.de (B.C.); maja.kahaly@googlemail.com (M.K.); thomas.kolben@med.uni-muenchen.de (T.K.); alexander.burges@med.uni-muenchen.de (A.B.); sven.mahner@med.uni-muenchen.de (S.M.); fabian.trillsch@med.uni-muenchen.de (F.T.)

[2] Institute of Pathology, Faculty of Medicine, 81377 LMU Munich, Germany; doris.mayr@med.uni-muenchen.de (D.M.); elisa.schmoeckel@med.uni-muenchen.de (E.S.)

[3] Institute of Human Genetics, Department of Human Molecular Genetics, University of Heidelberg, 69120 Heidelberg, Germany; Beate.Niesler@med.uni-heidelberg.de

* Correspondence: udo.jeschke@med.uni-muenchen.de; Tel.: +49-89-4400-74775

† These authors contributed equally to this work.

Received: 20 November 2018; Accepted: 21 December 2018; Published: 29 December 2018

Abstract: Nuclear factor erythroid 2-related factor 2 (NRF2) regulates cytoprotective antioxidant processes. In this study, the prognostic potential of NRF2 and its interactions with the estrogen receptor α (ERα) in ovarian cancer cells was investigated. NRF2 and ERα protein expression in ovarian cancer tissue was analyzed as well as mRNA expression of NRF2 (*NFE2L2*) and ERα (*ESR1*) in four ovarian cancer and one benign cell line. *NFE2L2* silencing was carried out to evaluate a potential interplay between NRF2 and ERα. Cytoplasmic NRF2 expression as inactive form had significantly higher expression in patients with low-grade histology ($p = 0.03$). In the serous cancer subtype, high cytoplasmic NRF2 expression (overall survival (OS), median 50.6 vs. 29.3 months; $p = 0.04$) and high ERα expression (OS, median 74.5 vs. 27.1 months; $p = 0.002$) was associated with longer overall survival as well as combined expression of both inactive cytoplasmic NRF2 and ERα in the whole cohort (median 74.5 vs. 37.7 months; $p = 0.04$). Cytoplasmic NRF2 expression showed a positive correlation with ERα expression ($p = 0.004$). *NFE2L2* was found to be highly expressed in the ovarian cancer cell lines OVCAR3, UWB1.289, and TOV112D. Compared with the benign cell line HOSEpiC, *ESR1* expression was reduced in all ovary cancer cell lines (all $p < 0.001$). Silencing of *NFE2L2* induced a higher mRNA expression of *ESR1* in the *NFE2L2* downregulated cancer cell lines OVCAR3 ($p = 0.003$) and ES2 ($p < 0.001$), confirming genetic interactions of NRF2 and ERα. In this study, both inactive cytoplasmic NRF2 and high ERα expression were demonstrated to be associated with improved survival in ovarian cancer patients. Further understanding of interactions within the estradiol–ERα–NRF2 pathway could better predict the impact of endocrine therapy in ovarian cancer.

Keywords: estrogen receptor alpha; nuclear factor erythroid 2-related factor 2; ovarian cancer; immunohistochemistry

---

1. Introduction

Ovarian cancer is the eighth most frequent cause of cancer death among women and the most lethal gynecological malignancy [1]. Relative five-year survival is less than 50% for patients with epithelial ovarian carcinoma (EOC) [2]. Main reasons for poor prognosis are insufficient screening methods, late stage detection, and resistance to chemotherapy later in the clinical course. As most patients have advanced stage disease, recommended therapy consists of cytoreductive surgery and platinum-based chemotherapy which might be combined with antiangiogenic bevacizumab. Residual disease after initial debulking surgery is the most important prognostic factor being influenced by treating physicians, while further clinical and pathological prognostic factors include the degree of differentiation, the International Federation of Gynecology and Obstetrics (FIGO) stage, and histological subtype [3–6]. With serous, mucinous, endometrioid, and clear cell histology, invasive EOC exhibits several histopathological subtypes that are phenotypically, molecularly, and etiologically distinct [7]. The association between tumor biomarker expression and survival varies substantially between subtypes and can be distinguished in overall analyses of all EOCs [8,9].

According to current investigations, the occurrence of EOC seems to be related to oxidative stress [9]. By activating the nuclear factor erythroid-2-related factor 2 (NRF2), a relevant regulator of antioxidant and cytoprotective genes, both healthy and tumor cells can cope with oxidative stress. NRF2 is ubiquitously expressed at low levels in all human organs. As NRF2 regulates a major cellular defense mechanism, tight regulation is crucial to maintain cellular homeostasis. High constitutive levels of NRF2 have been described in different tumors or cancer cell lines [10–14]. Overexpression of NRF2 might protect cancer cells from the cytotoxic effects of anticancer therapies, resulting in resistance for chemo- or radiotherapy [15,16].

So far, the role of estrogen in EOC is still debated [17]. While application of exogenous hormones for menopause-related symptoms could be associated with an increased risk of EOC [18], a protective effect of oral contraceptives has been described. The estrogen receptor (ER) is expressed in two isoforms, the ERα and ERß [19]. ERα mediates the effects of female steroid hormones on proliferation and apoptosis of EOC cells, and immunohistochemical assessment of ER status is routinely done for the clinical management of breast cancer [19]. Molecular and cell biological interactions between NRF2 and ERα have been reported so far [16,20]. The aryl hydrocarbon receptor and ERα differentially modulate NRF2 transactivation in MCF-7 breast cancer cells [16]. Furthermore, studies show an important crosstalk between NRF2 and ERα in neurophysiological processes [16,20].

To better understand these effects in EOC, we first assessed the prognostic influence of NRF2 and ERα in various subtypes of EOC. To understand the interaction of NRF2 and Erα on a molecular level, we investigated the expression and their correlation in vitro.

2. Results

2.1. NRF2/ERα Expression Correlates with Cinical and Pathological Data

Nuclear staining of NRF2 was technically successful in 145 of 156 cases (93%) with positive staining in 144 of 145 cases (99%). Cytoplasmic staining of NRF2 was evaluable with technically adequate staining in 139 of 156 cases (89%) (Figure 1 and Figures S1 and S2) and NRF2 expression was observed in all these 139 specimens (100%). Median (range) immunoreactivity scores (IRS) for NRF2 in nuclei and cytoplasm were 8 (2,12) and 8 (4,12), respectively.

NRF2 expression displayed correlations to clinical and pathological data (Table 1). NRF2 staining in both cytoplasm and nucleus was different between the histological subtypes ($p = 0.001$ and $p = 0.02$, respectively) with low nuclear NRF2 expression in serous, clear cell, and endometrioid histology and high expression in the mucinous subtype. In comparison, the strongest and weakest cytoplasmic NRF2 staining was found in the serous and clear cell subtypes, respectively. Cytoplasmic NRF2 expression had significantly higher expression in patients with low-grade histology ($p = 0.03$), and low nuclear NRF2 expression was associated with age ($p = 0.045$) (Table 1).

ERα staining was successfully performed in all 156 cases (100%), and ERα expression was observed in 70 of 156 (45%) specimens with a median (range) IRS of 4 (1,12) (Figure 1 and Figure S1). There was no significant difference in the ERα expression comparing all histological subtypes ($p = 0.21$). Analysis of clear cell and endometrioid ovarian cancer subtypes revealed nearly significant upregulation ($p = 0.05$). Analyzing the grading, there were no significant differences in general, and low-graded patients showed significantly higher ERα expression compared to high-graded patients

(p = 0.028). All other parameters, such as FIGO, lymph node involvement (pN), and distant metastasis (pM), showed no significant differences in the ERα expression. NRF2 cytoplasmic expression correlated with ERα expression (p = 0.004, Table 2 and Figures 1 and 2).

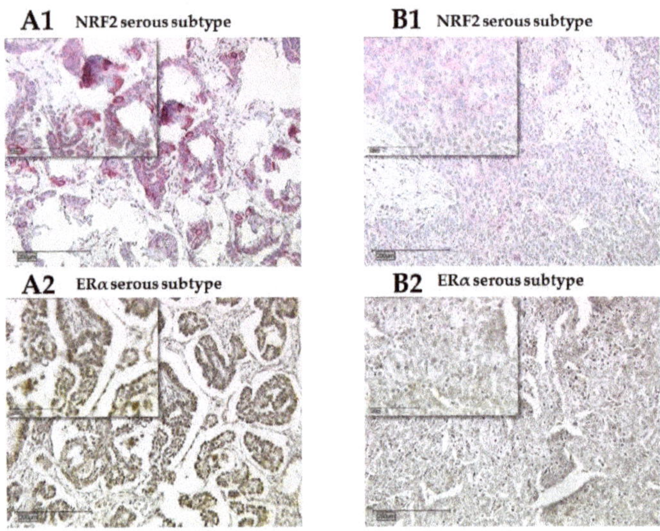

Figure 1. Detection of nuclear factor erythroid-2-related factor 2 (NRF2) (**A1, B1**) and estrogen receptor (ER)α (**A2, B2**) with immunohistochemistry. High (**A1**) and low (**B1**) cytoplasmic NRF2 stains in serous subtype correspond with high (**A2**) and low (**B2**) ERα stains, respectively. NRF2 shows faint staining in the nucleus in both cases (**A1, B1**).

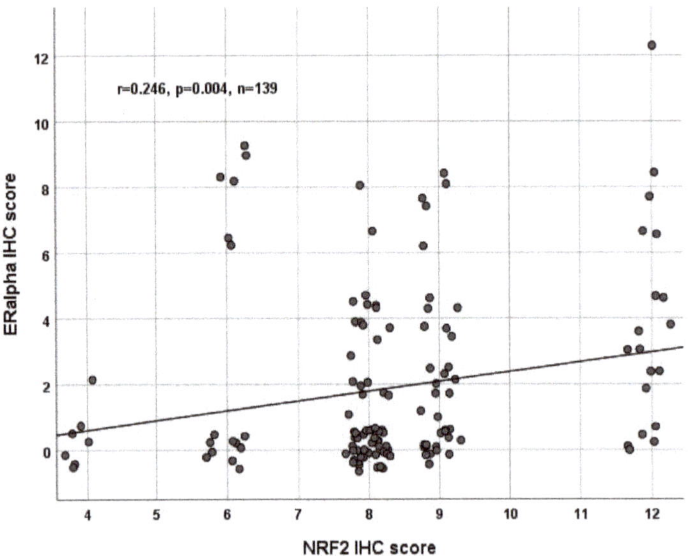

Figure 2. Correlation analysis of NRF2 and ERα in ovarian cancer tissue (n = 139). A significant correlation of cytoplasmic NRF2 expression with ERα expression was noted. For better visualization, dots have been jittered.

Table 1. Expression profile of NRF2 staining regarding clinical and pathological characteristics.

Parameters	N	Nuclear NRF2 Expression			p	N	Cytoplasmic NRF2 Expression			p
		Negative	Low	High			Negative	Low	High	
Histology										
Serous	103	0	87	16	0.02	98	0	54	44	0.001
Clear cell	11	1	7	3		11	0	11	0	
Endometrioid	20	0	18	2		19	0	12	7	
Mucinous	11	0	3	8		11	0	7	4	
Lymph node										
pN0/X	96	0	76	20	NS	93	0	59	34	NS
pN1	49	1	39	9		46	0	25	21	
Distant Metastasis										
pM0/X	141	1	112	28	NS	135	0	83	52	NS
pM1	4	0	3	1		4	0	1	3	
Grading										
Low	33	0	25	8	NS	33	0	16	17	0.03
High	100	1	83	16		95	0	64	31	
FIGO										
I/II	41	0	31	10	NS	40	0	24	16	NS
III/IV	99	0	81	18		94	0	56	38	
Age										
≤60 years	77	1	56	20	0.045	75	0	43	32	NS
>60 years	68	0	59	9		64	0	41	23	

Table 2. Correlation analysis.

Staining	NRF2 Nucleus	NRF2 Cytoplasm	ERα
NRF2 Nucleus			
cc	1.000	0.013	−0.019
p		0.88	0.82
n	146	138	146
NRF2 Cytoplasm			
cc	0.013	1.000	0.246
p	0.88		0.004
n	138	139	139
ERα			
cc	−0.019	0.246	1.000
p	0.82	0.004	
n	146	139	156

Immunoreactivity scores (IRS) of NRF2 and ERα staining in different compartments was correlated to each other using Spearman's correlation analysis. cc = correlation coefficient, p = two-tailed significance, n = number of patients.

2.2. High NRF2/ERα Expression is Associated with Improved Overall Survival

The median age of the patients was 58.7 (standard deviation (SD) of 31.4) years with a range of 31–88 years. Median overall survival of the EOC patients was 34.4 (SD 57.8) months. Cytoplasmic NRF2 expression in the serous cancer subtype was associated with longer overall survival (Figure 3, median 50.6 vs. 29.3 months; $p = 0.04$) as it was noted for ERα expression (Figure 3, median 74.5 vs. 27.1 months; $p = 0.002$). Improved OS was also seen for patients with combined and high expression of both NRF2 and ERα in the cytoplasm comparing all histological subtypes (Figure 3, median 74.5 vs. 37.7 months; $p = 0.04$).

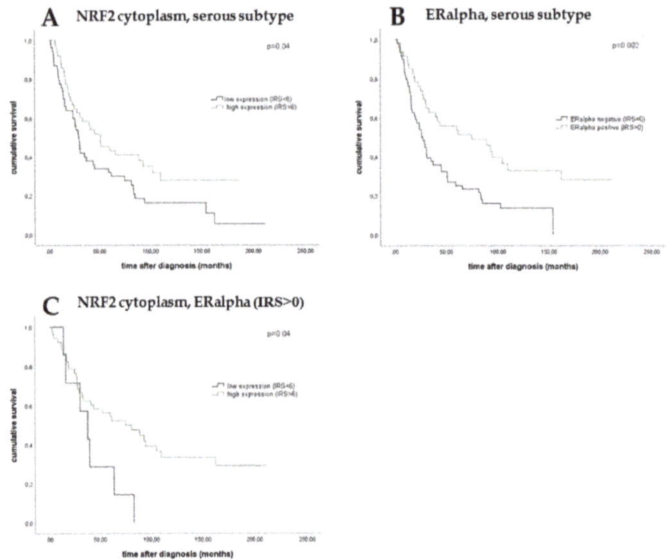

Figure 3. Kaplan–Meier estimates of NRF2 expression, ERα expression, and combined NRF2 and ERα expression were analyzed. In the serous subtype, patients with a high cytoplasmic expression of NRF2 showed a significantly increased overall survival compared with patients with a low cytoplasmic expression (**A**). In addition, high ERα expression was associated with significantly better overall survival in serous ovarian cancer compared with patients with a low ERα expression (**B**). Patients with combined high NRF2 expression in the cytoplasm and ERα expression in epithelial ovarian carcinoma (EOC) had significantly increased overall survival compared with those with low cytoplasmic expression and ERα expression (**C**).

2.3. Clinical and Pathological Parameters are Independent Prognostic Factors

Cancer grading, the FIGO classification, and patients' age were independent prognostic factors in the present cohort (Table 3). In contrast, prognostic impact of histological subtype, NRF2, and/or ERα staining/expression was not significant.

Table 3. Multivariate analysis.

Covariate	Coefficient (b_i)	[HR Exp(b_i)]	95% CI		p-Value
			Lower	Upper	
Histology (serous vs. other)	−0.108	0.898	0.678	1.188	0.45
Grade (low vs. high)	0.519	1.680	1.211	2.332	0.002
FIGO (I, II vs. III, IV)	0.722	2.058	1.421	2.979	0.000
Patients' age ($\leq$60 vs. >60 years)	0.000	1.000	1.000	1.000	0.001
NRF2 cytoplasmic/ ERα	−0.166	0.847	0.531	1.351	0.49

2.4. Downregulation of NFE2L2 Increases ESR1 Expression, Confirming Their Genetic Interaction

Basal expressions of both *NFE2L2* and *ESR1* were analyzed by qPCR in all four EOC cell lines and compared with a benign ovarian cell line (HOSEpiC). As shown in Figure 4 and compared to HOSEpiC, *NFE2L2* expression increased 2-fold in both OVCAR3 ($p = 0.02$) and UWB1.289 ($p = 0.08$)

and was 1.5-fold elevated in the TOV112D ($p = 0.30$) cell lines. In comparison, *ESR1* expression was markedly reduced in all EOC cell lines compared to the benign ovarian cells (all $p < 0.001$).

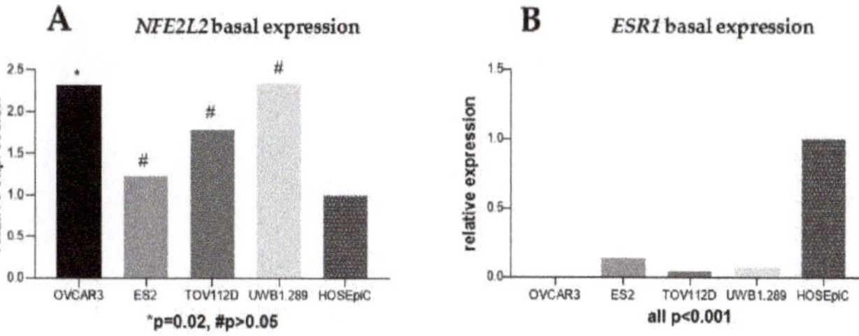

Figure 4. Basal gene expression of *NFE2L2* (**A**) and *ESR1* (**B**) in four ovarian cancer cell lines was compared to the expression in the benign ovarian cell line (HOSEpiC).

Following effective silencing of *NFE2L2* with siRNA to evaluate the impact on *ESR1* expression (Figure 5), an elevated expression of *ESR1* in the *NFE2L2* downregulated cancer cell lines OVCAR3 ($p = 0.003$) and ES2 was noted ($p < 0.001$).

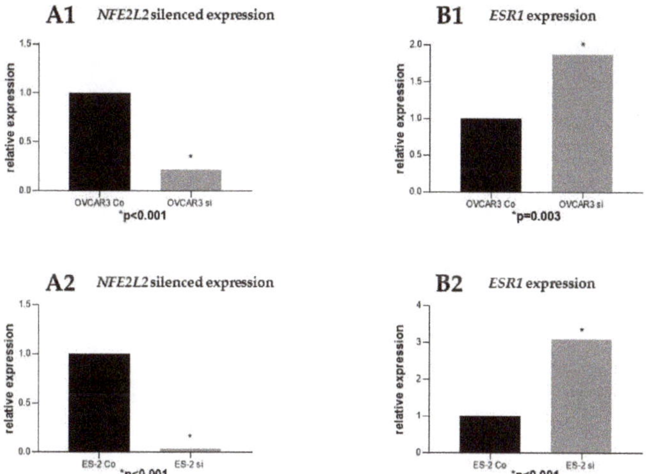

Figure 5. siRNA downregulation of *NFE2L2* in the ovarian cancer cell lines OVCAR3 (**A1**) and ES2 (**A2**). *ESR1* expression following *NFE2L2* downregulation in both cell lines (**B1, B2**).

3. Discussion

This cell and molecular biological experimental study reveals that NRF2 expression differs in histologic subtypes of EOC, with the strongest cytoplasmic expression in the serous subtype. Cytoplasmic NRF2 expression had significantly higher expression in patients with low-grade histology. Patients with higher cytoplasmic NRF2 expression in the serous type confirmed to have a significantly improved OS. Moreover, we could reveal that the combination of cytoplasmic NRF2 and ERα expression was associated with significantly longer OS. Molecular testing in cell lines exhibited that the *ESR1* gene was lower expressed in all four EOC cell lines, which could be upregulated by *NFE2L2* silencing in the subsequently *NFE2L2*-downregulated cancer cell lines.

NRF2 has been traditionally considered as a tumor suppressor because its cytoprotective functions are deemed to be the main cellular defense mechanism against exogenous and endogenous insults, including xenobiotic and oxidative stress [21,22]. Under homeostatic conditions, NRF2 activation prevents excessive cellular damage produced by metabolic, xenobiotic, and oxidative stress [22]. NRF2 activation is thus important in cancer chemoprevention. Cancer chemoprevention mechanisms seem to be mediated through the Keap1–NRF2 pathway, and in experimental models, NRF2/Keap1 mutations are present at preneoplastic stages [23]. Further, NRF2-null mice are more prone to develop cancer in response to chemical and physical stimuli (nitrosamine, ultraviolet light, and aflatoxin) [17]. On the other hand, recent studies demonstrated that NRF2 hyperactivation may also create an environment favoring survival of normal as well as malignant cells, protecting them from apoptosis and senescence and against oxidative stress, chemotherapeutic agents, and radiotherapy [24,25]. Hence, the potential dual role of NRF2 in cancer may explain the described results below.

Our findings are in line with previous reports showing that nuclear or activated NRF2 expression is associated with upregulation of NRF2 target genes and poorer OS and disease-free survival (DFS), whereas patients with high cytoplasmic or inactive NRF2 expression displayed better OS and DFS [26]. Our evaluation of ER expression in the EOC tissue samples confirmed previous reports. In patients with EOC, the ER, especially ERα, is significantly associated with improved OS [8], grading, progression-free survival, and cause-specific survival, respectively [27]. There is a strong relationship between circulating sex hormones and female reproductive cancers (e.g., ovarian, breast, and endometrial cancers) [28]. Interestingly, estradiol may play a dual role in modulating NRF2 activity. On the one hand, its metabolites activate NRF2 via the generation of reactive oxygen species (ROS) (independent of the ER) [29]. While recent studies demonstrated that estradiol leads to an activation of NRF2 in a wide range of cell types [30,31], the estradiol effect was only noted on protein and not on mRNA levels, suggesting that the main effect of estradiol is based on NRF2 protein stabilization [32]. However, binding to ERα (dependent of ER) appears to be the mechanism for estradiol itself to inhibit the NRF2 downstream genes [9]. ERα, but not ERβ, interacts with NRF2 in an estradiol-dependent way and thereby represses NRF2-mediated transcription [33]. Thus, EOC patients with high tumor expression of ERα show a strong influence of the estradiol–ERα-dependent pathway, resulting in inactivated NRF2 and better survival rates. Otherwise, low ERα expression causes a dysbalance in favor of the estradiol–ERα-independent pathway with an activation of NRF2 (Figure 6). Studies show that other NRF2-associated factors also could play a crucial role in the above-described interaction. Glutathione S-transferase (GST), an NRF2 target gene, is modulated by miR-186 overexpression in OVCAR3 cells with consecutively increased sensitivity of ovarian cancer cells to paclitaxel [34]. Furthermore, it was described that the KEAP1–NRF2 pathway is important in ovarian cancer cell reaction to cigarette-smoke-induced ROS [35].

Endocrine therapy in EOC has been considered as a potential approach in subgroups of patients with a specific tumor biology that responds to this therapy [36]. Hereby, the rationale for endocrine treatment is based on the high ER/PR IHC expression as a predictive marker [37]. A present prospective study demonstrated evidence for the usefulness of letrozole as an aromatase inhibitor in serous EOC [38]. Under the conditions described above, treatment with aromatase inhibitors could cause a prognostically beneficial predominance of the ERα–NRF2-dependent pathway. As revealed in the present investigation, a putative functional association of endocrine therapy and NRF2 underlines the relationship of NRF2/ERα, as confirmed by significant correlation of expression. In addition to the mentioned approach, further therapeutic strategies as interference of DNA repair mechanisms are of great interest to overcome treatment burden [39–42].

The retrospective design, the relatively small number of tissue samples evaluated, and the semiquantitative scoring method may critically be regarded as limitations of the submitted work. The data are hypothesis generating and further prospective studies with a larger patient collective and standardized immunohistochemical and molecular methods are warranted to gain more detailed and better insight into this research field.

However, despite these drawbacks, our analysis indicates for the first time a putative molecular role of the estradiol–ERα–NRF2 pathway as a basis for a better understanding of endocrine therapy in EOC.

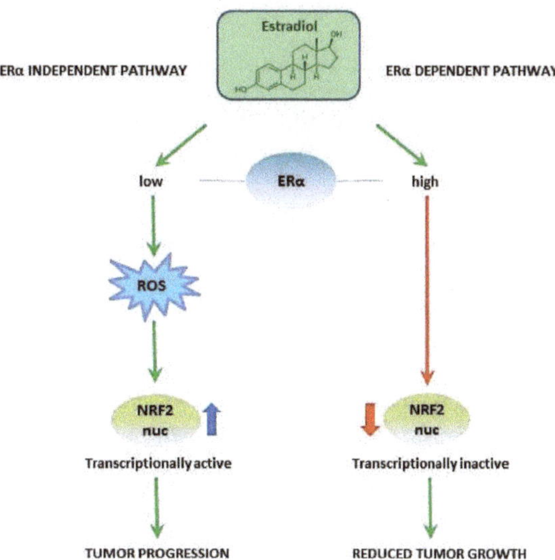

Figure 6. Summary of the hypothesized interaction within the estradiol–ERα–NRF2 pathway: High expression of ERα leads to an induction of the estradiol–ERα-dependent pathway, resulting in transcriptionally inactive NRF2 (low nuclear, high cytoplasmic expression) and consecutively less impact on tumor growth. In contrast, low ERα expression favors the estradiol–ERα-independent pathway, with activation of NRF2 (high nuclear, low cytoplasmic expression) causing tumor progression.

4. Materials and Methods

4.1. Ethical Approval

The current study was approved by the Ethics Committee of the Ludwig-Maximilians-University, Munich, Germany (approval number 227-09) on 30 September 2009. All tissue samples used for this study were obtained from material from the archives of LMU Munich, Department Gynecology and Obstetrics, Ludwig-Maximilians-University, Munich, Germany, initially used for pathological diagnostics. The diagnostic procedures were completed before the current study was performed. During the analysis, the observers were fully blinded to patients' data. The study was approved by the Ethics Committee of LMU Munich. All experiments were performed according to the standards of the Declaration of Helsinki (1975).

4.2. Patients and Specimens

Tissue samples of 156 patients who underwent surgery for EOC at the Department of Obstetrics and Gynecology, Ludwig-Maximillian's-University Munich from 1990 to 2002 were analyzed in this study. Clinical data was obtained from the patients' charts and follow up data from the Munich Cancer Registry. All samples had been formalin-fixated and paraffin-embedded (FFPE). Patients with benign or borderline tumors were excluded and no patients had adjuvant chemotherapy. Specialized pathologists for EOC examined and classified the samples for tumor grading—low ($n = 38$), high ($n = 117$)—and histological subtypes—serous ($n = 110$), endometrioid ($n = 21$), clear cell ($n = 12$), and mucinous ($n = 13$). Staging was performed using TNM and FIGO (WHO) classification: I ($n = 35$), II ($n = 10$,) III ($n = 103$), and IV ($n = 3$). Data on primary tumor extension were available in 155 cases—T1 ($n = 40$), T2 ($n = 18$), T3 ($n = 93$), and T4 ($n = 4$)—as well as data on lymph node involvement in 95 cases—N0 ($n = 43$), N1 ($n = 52$). Data on distant metastasis were available in nine cases—M0 ($n = 3$), M1 ($n = 6$).

4.3. Immunohistochemistry

Immunohistochemistry was performed as previously described by our lab [43]. For NRF2 staining, FFPE EOC samples were incubated with anti-NRF2 (Abcam, Cambridge, UK, rabbit, monoclonal, clone EP1808) at a final concentration of 5.93 µg/ml (1:100 dilution) for 1 h at room temperature. Afterwards, slides were incubated with isotype-matching MACH 3 Rabbit AP Polymer Detection (Biocare Medical, Pacheco, CA, USA, catalogue-number M3R533). The Permanent AP Red Kit (Zytomed Systems GmbH, Berlin, Germany, catalogue-number ZUC-001) was used as a chromogen. Slides were then counterstained with Gill's hematoxylin (Vector Laboratories, Burlingame, CA, USA). System controls were included.

For the detection of ERα, resected EOC tissue samples were fixed in formalin and embedded in paraffin after surgery. ERα staining was performed by blocking slides with goat serum (1:100 dilution, Vectastain® ABC-Elite-Kit, Linaris, Dossenheim, Germany, catalogue-number PK-6101) for 30 min at room temperature. Subsequently, slides were incubated with anti-ERα primary antibody (1:400 dilutions, Abcam, Cambridge, UK, rabbit, monoclonal, clone EPR703(2)) for 16 h at 4 °C. Afterwards, slides were incubated with isotype-matching anti-rabbit IgG secondary antibody and avidin–biotin–peroxidase complex both for 30 min at room temperature, according to the Vectastain® ABC-Elite-Kit (Linaris, Dossenheim, Germany, catalogue-number PK-6101). All slides were washed twice in PBS for 2 min after every incubation step. 3,3'-Diaminobenzidine chromogen (DAB; Dako, Glostrup, Denmark, catalogue-number K3468) was used for visualization reaction. Slides were then counterstained with Mayer's acidic hematoxylin (Waldeck-Chroma, Münster, Germany, catalogue number 2E-038) and dehydrated in an ascending series of alcohol followed by xylol. System controls were included.

4.4. Staining Evaluation and Statistical Analysis

All EOC specimens were examined with a Leitz (Wetzlar, Germany) photomicroscope and specific NRF2 and ERα immunohistochemical staining reaction was observed in the nuclei and cytoplasm of the cells. The intensity and distribution pattern of NRF2 and ERα staining was rated using the semiquantitative immunoreactivity score (IRS, Remmele's score). To obtain the IRS result, the optional staining intensity (0 = no, 1 = weak, 2 = moderate, and 3 = strong staining) and the percentage of positive stained cells (0 = no staining, 1 = <10% of the cells, 2 = 11%–50% of the cells, 3 = 51%–80% of the cells, and 4 = >81%) were multiplied. Nuclear and cytoplasmic NRF2 staining was successfully performed in 145 (93%) and 139 (89%) of 156 EOC tissue specimens, respectively. Cut-off points for the IRSs were selected for cytoplasmic and nuclear NRF2 staining considering the distribution pattern of IRSs in the collective. Nuclear and cytoplasmic NRF2 staining were regarded as negative with an IRS of 0–2, as low with IRS of 4–8, and as high with IRS of $\geq$9. ERα staining was successfully performed in all 156 (100%) EOC specimens. Cellular ERα staining was considered as negative with an IRS of 0 and as positive with an IRS of >0.

Statistical analysis was performed using SPSS 25.0 (v25, IBM, Armonk, New York). Distribution of clinical pathological variables was evaluated with the chi-squared test. The Mann–Whitney *U* test was used to compare IRSs of NRF2 between different clinical and pathological subgroups. Correlations between findings of immunohistochemical staining were calculated using Spearman's analysis. Survival times were analyzed by Kaplan–Meier (log-rank) estimates. To identify an appropriate cut-off, the ROC curve was drawn, which is considered as one of the most reliable methods for cut-off point selection. In this context, the ROC curve was a plot representing sensitivity on the y-axis and (1-specificity) on x-axis [44]. Consecutively, Youden's index, defined as the maximum (sensitivity+specificity-1) [45], was used to find the optimal cut-off maximizing the sum of sensitivity and specificity [46,47]. For multivariate analyses, a Cox regression model was applied, with *p*-values less than 0.05 considered to be significant. Ct values of each gene were obtained with qPCR and the relative expressions were calculated using the $2^{-\Delta\Delta Ct}$ formula. Statistical data was acquired using Graph Pad Prism 7.03 (v7, La Jolla, CA, USA).

4.5. Cell Lines

The human ovarian cancer cell lines OVCAR3 (serous), ES-2 (clear cell), TOV112D (endometrioid), and UWB1.289 (serous, BRCA1 negative) were purchased from the American Type Culture Collection (ATCC, Rockville, MD, USA). Cells were maintained in culture in RPMI 1640 medium (ThermoFisher Scientific, Waltham, MA, USA) supplemented with 10% FBS in a humidified incubator at 37 °C under 5% CO_2. The benign ovarian cell line HOSEpiC was purchased from ScienCell (Carlsbad, CA, USA). HOSEpiC cells were maintained in culture in Ovarian Epithelial Cell Medium (OEpiCM) (ScienCell, Carlsbad, CA, USA, catalogue-number 7311) in a humidified incubator at 37 °C under 5% CO_2.

4.6. PCR

RNA isolation was performed using the RNeasy Mini Kit (Qiagen, Venlo, Netherlands) and 1 µg of RNA was converted into first-strand cDNA using the MMLV Reverse Transcriptase 1st-Strand cDNA Synthesis Kit (Epicentre, Madison, WI, USA) according to the instructions of the manufacturer. The basal mRNA expressions of *NFE2L2* and *ESR1* were quantified by qPCR applying FastStart Essential DNA Probes Master and gene-specific primers (Roche, Basel, Switzerland). For normalization of expressions the housekeeping genes, *β-Actin* and *GAPDH* were used as reference controls. Basal expressions of *NFE2L2* and *ESR1* in the ovarian cancer cell lines were compared with their expressions in the benign ovarian cell lines.

4.7. siRNA

The specific siRNA for *NFE2L2* (Silencer Select Pre-designed and Custom Designed siRNA, Ambion, Carlsbad, CA, USA) was kindly provided by Beate Niesler (Department of Human Molecular Genetics, University of Heidelberg). Cells were transfected with siRNA using Lipofectamine RNAiMAX reagent (Invitrogen, Carlsbad, CA, USA) to silence the expression of *NFE2L2* in the cell lines. RNA isolation and mRNA quantification by qPCR was repeated as outlined above. mRNA expression levels of *NFE2L2* and *ESR1* in *NFE2L2*-downregulated cells were compared with *NFE2L2*-containing cells.

5. Conclusions

Here, *ESR1* expression was reduced in different ovarian cancer cells vs. benign cells in vitro (all $p < 0.001$). *NFE2L2* silencing showed a higher expression of *ESR1* in the *NFE2L2*-downregulated cancer cell lines OVCAR3 ($p = 0.003$) and ES2 ($p < 0.001$). In the serous cancer subtype, high cytoplasmic NRF2 expression (OS, median 50.6 vs. 29.3 months; $p = 0.04$) and high ERα expression (OS, median 74.5 vs. 27.1 months; $p = 0.002$) was associated with longer overall survival as well as combined expression of both inactive cytoplasmic NRF2 and ERα in the whole cohort (median 74.5 vs. 37.7 months; $p = 0.04$). Thus, interactions of NRF2 and ERα impact survival in ovarian cancer patients and may be important factors for the response to endocrine treatment strategies.

Supplementary Materials: Supplementary materials can be found at http://www.mdpi.com/1422-0067/20/1/112/s1.

Author Contributions: B.C. and U.J. conceived and designed the experiments; M.K. performed the experiments; M.K. and U.J. analyzed the data; D.M., E.S., and B.N. contributed reagents/materials/analysis tools; B.C., M.K., and F.T. wrote the paper. D.M., E.S., B.N., T.K., A.B., S.M., and U.J. critically reviewed the paper.

Funding: This work has been funded by the "Monika Kutzner" foundation and the "Brigitte & Dr. Konstanze Wegener" foundation.

Acknowledgments: The authors are grateful to Martina Rahmeh and Christina Kuhn for excellent technical assistance.

Conflicts of Interest: Bastian Czogalla has received a research grant from the "Monika Kutzner" foundation and "Brigitte & Dr. Konstanze Wegener" foundation, all other authors declare no conflict of interest.

References

1. Prat, J.; Franceschi, S. Cancers of the female reproductive organs. In *World Cancer Report*; Stewart, B.W., Wild, C.P., Eds.; IARC: Lyon, France, 2014; pp. 465–481.
2. Baldwin, L.A.; Huang, B.; Miller, R.W.; Tucker, T.; Goodrich, S.T.; Podzielinski, I.; DeSimone, C.P.; Ueland, F.R.; van Nagell, J.R.; Seamon, L.G. Ten-year relative survival for epithelial ovarian cancer. *Obstet. Gynecol.* **2012**, *120*, 612–618. [CrossRef]
3. du Bois, A.; Reuss, A.; Pujade-Lauraine, E.; Harter, P.; Ray-Coquard, I.; Pfisterer, J. Role of surgical outcome as prognostic factor in advanced epithelial ovarian cancer: A combined exploratory analysis of 3 prospectively randomized phase 3 multicenter trials: By the Arbeitsgemeinschaft Gynaekologische Onkologie Studiengruppe Ovarialkarzinom (AGO-OVAR) and the Groupe d'Investigateurs Nationaux Pour les Etudes des Cancers de l'Ovaire (GINECO). *Cancer* **2009**, *115*, 1234–1244.
4. Aletti, G.D.; Gostout, B.S.; Podratz, K.C.; Cliby, W.A. Ovarian cancer surgical resectability: RELATIVE impact of disease, patient status, and surgeon. *Gynecol. Oncol.* **2006**, *100*, 33–37. [CrossRef]
5. Vergote, I.; De Brabanter, J.; Fyles, A.; Bertelsen, K.; Einhorn, N.; Sevelda, P.; Gore, M.E.; Kærn, J.; Verrelst, H.; Sjövall, K.; et al. Prognostic importance of degree of differentiation and cyst rupture in stage I invasive epithelial ovarian carcinoma. *Lancet* **2001**, *357*, 176–182. [CrossRef]
6. Dembo, A.J.; Davy, M.; Stenwig, A.E.; Berle, E.J.; Bush, R.S.; Kjorstad, K. Prognostic factors in patients with stage I epithelial ovarian cancer. *Obstet. Gynecol.* **1990**, *75*, 263–273.
7. Kossai, M.; Leary, A.; Scoazec, J.Y.; Genestie, C. Ovarian Cancer: A Heterogeneous Disease. *Pathobiol. J. Immunopathol. Mol. Cell. Boil.* **2018**, *85*, 41–49. [CrossRef]
8. Sieh, W.; Kobel, M.; Longacre, T.A.; Bowtell, D.D.; deFazio, A.; Goodman, M.T.; Hogdall, E.; Deen, S.; Wentzensen, N.; Moysich, K.B.; et al. Hormone-receptor expression and ovarian cancer survival: An Ovarian Tumor Tissue Analysis consortium study. *Lancet Oncol.* **2013**, *14*, 853–862. [CrossRef]
9. van der Wijst, M.G.; Brown, R.; Rots, M.G. Nrf2, the master redox switch: The Achilles' heel of ovarian cancer? *Biochim. Biophys. Acta* **2014**, *1846*, 494–509. [CrossRef]
10. Namani, A.; Matiur Rahaman, M.; Chen, M.; Tang, X. Gene-expression signature regulated by the KEAP1-NRF2-CUL3 axis is associated with a poor prognosis in head and neck squamous cell cancer. *BMC Cancer* **2018**, *18*, 46. [CrossRef]
11. Boustani, M.R.; Khoshnood, R.J.; Nikpasand, F.; Taleshi, Z.; Ahmadi, K.; Yahaghi, E.; Goudarzi, P.K. Overexpression of ubiquitin-specific protease 2a (USP2a) and nuclear factor erythroid 2-related factor 2 (Nrf2) in human gliomas. *J. Neurol. Sci.* **2016**, *363*, 249–252. [CrossRef]
12. Ji, L.; Wei, Y.; Jiang, T.; Wang, S. Correlation of Nrf2, NQO1, MRP1, cmyc and p53 in colorectal cancer and their relationships to clinicopathologic features and survival. *Int. J. Clin. Exp. Pathol.* **2014**, *7*, 1124–1131. [PubMed]
13. Jiang, T.; Chen, N.; Zhao, F.; Wang, X.J.; Kong, B.; Zheng, W.; Zhang, D.D. High levels of Nrf2 determine chemoresistance in type II endometrial cancer. *Cancer Res.* **2010**, *70*, 5486–5496. [CrossRef] [PubMed]
14. Ryoo, I.G.; Choi, B.H.; Kwak, M.K. Activation of NRF2 by p62 and proteasome reduction in sphere-forming breast carcinoma cells. *Oncotarget* **2015**, *6*, 8167–8184. [CrossRef] [PubMed]
15. Zhang, D.D. The Nrf2-Keap1-ARE signaling pathway: The regulation and dual function of Nrf2 in cancer. *Antioxid. Redox Signal.* **2010**, *13*, 1623–1626. [CrossRef]
16. Lo, R.; Matthews, J. The aryl hydrocarbon receptor and estrogen receptor alpha differentially modulate nuclear factor erythroid-2-related factor 2 transactivation in MCF-7 breast cancer cells. *Toxicol. Appl. Pharmacol.* **2013**, *270*, 139–148. [CrossRef]
17. Lenhard, M.; Tereza, L.; Heublein, S.; Ditsch, N.; Himsl, I.; Mayr, D.; Friese, K.; Jeschke, U. Steroid hormone receptor expression in ovarian cancer: Progesterone receptor B as prognostic marker for patient survival. *BMC Cancer* **2012**, *12*, 553. [CrossRef]
18. Lukanova, A.; Kaaks, R. Endogenous hormones and ovarian cancer: Epidemiology and current hypotheses. *Cancer Epidemiol. Biomark. Prev.* **2005**, *14*, 98–107.
19. Cunat, S.; Hoffmann, P.; Pujol, P. Estrogens and epithelial ovarian cancer. *Gynecol. Oncol.* **2004**, *94*, 25–32. [CrossRef]

20. Zhu, C.; Wang, S.; Wang, B.; Du, F.; Hu, C.; Li, H.; Feng, Y.; Zhu, R.; Mo, M.; Cao, Y.; et al. 17beta-Estradiol up-regulates Nrf2 via PI3K/AKT and estrogen receptor signaling pathways to suppress light-induced degeneration in rat retina. *Neuroscience* **2015**, *304*, 328–339. [CrossRef]

21. Villeneuve, N.F.; Lau, A.; Zhang, D.D. Regulation of the Nrf2-Keap1 antioxidant response by the ubiquitin proteasome system: AN insight into cullin-ring ubiquitin ligases. *Antioxid. Redox Signal.* **2010**, *13*, 1699–1712. [CrossRef]

22. Wakabayashi, N.; Slocum, S.L.; Skoko, J.J.; Shin, S.; Kensler, T.W. When NRF2 talks, who's listening? *Antioxid. Redox Signal.* **2010**, *13*, 1649–1663. [CrossRef] [PubMed]

23. Hayes, J.D.; McMahon, M.; Chowdhry, S.; Dinkova-Kostova, A.T. Cancer chemoprevention mechanisms mediated through the Keap1-Nrf2 pathway. *Antioxid. Redox Signal.* **2010**, *13*, 1713–1748. [CrossRef] [PubMed]

24. Menegon, S.; Columbano, A.; Giordano, S. The Dual Roles of NRF2 in Cancer. *Trends Mol. Med.* **2016**, *22*, 578–593. [CrossRef] [PubMed]

25. Taguchi, K.; Yamamoto, M. The KEAP1-NRF2 System in Cancer. *Front. Oncol.* **2017**, *7*, 85. [CrossRef] [PubMed]

26. Cho, H.-Y.; Kim, K.; Kim, Y.-B.; Kim, H.; No, J.H. Expression Patterns of Nrf2 and Keap1 in Ovarian Cancer Cells and their Prognostic Role in Disease Recurrence and Patient Survival. *Int. J. Gynecol. Cancer* **2017**, *27*, 412–419. [CrossRef] [PubMed]

27. Burges, A.; Bruning, A.; Dannenmann, C.; Blankenstein, T.; Jeschke, U.; Shabani, N.; Friese, K.; Mylonas, I. Prognostic significance of estrogen receptor alpha and beta expression in human serous carcinomas of the ovary. *Arch. Gynecol. Obstet.* **2010**, *281*, 511–517. [CrossRef] [PubMed]

28. Chuffa, L.G.; Lupi-Junior, L.A.; Costa, A.B.; Amorim, J.P.; Seiva, F.R. The role of sex hormones and steroid receptors on female reproductive cancers. *Steroids* **2017**, *118*, 93–108. [CrossRef]

29. Liao, H.; Zhou, Q.; Zhang, Z.; Wang, Q.; Sun, Y.; Yi, X.; Feng, Y. NRF2 is overexpressed in ovarian epithelial carcinoma and is regulated by gonadotrophin and sex-steroid hormones. *Oncol. Rep.* **2012**, *27*, 1918–1924.

30. Gorrini, C.; Baniasadi, P.S.; Harris, I.S.; Silvester, J.; Inoue, S.; Snow, B.; Joshi, P.A.; Wakeham, A.; Molyneux, S.D.; Martin, B.; et al. BRCA1 interacts with Nrf2 to regulate antioxidant signaling and cell survival. *J. Exp. Med.* **2013**, *210*, 1529–1544. [CrossRef]

31. Gorrini, C.; Gang, B.P.; Bassi, C.; Wakeham, A.; Baniasadi, S.P.; Hao, Z.; Li, W.Y.; Cescon, D.W.; Li, Y.T.; Molyneux, S.; et al. Estrogen controls the survival of BRCA1-deficient cells via a PI3K-NRF2-regulated pathway. *Proc. Natl. Acad. Sci. USA* **2014**, *111*, 4472–4477. [CrossRef]

32. Symonds, D.A.; Merchenthaler, I.; Flaws, J.A. Methoxychlor and estradiol induce oxidative stress DNA damage in the mouse ovarian surface epithelium. *Toxicol. Sci.* **2008**, *105*, 182–187. [CrossRef] [PubMed]

33. Ansell, P.J.; Lo, S.C.; Newton, L.G.; Espinosa-Nicholas, C.; Zhang, D.D.; Liu, J.H.; Hannink, M.; Lubahn, D.B. Repression of cancer protective genes by 17beta-estradiol: Ligand-dependent interaction between human Nrf2 and estrogen receptor alpha. *Mol. Cell. Endocrinol.* **2005**, *243*, 27–34. [CrossRef] [PubMed]

34. Sun, K.X.; Jiao, J.W.; Chen, S.; Liu, B.L.; Zhao, Y. MicroRNA-186 induces sensitivity of ovarian cancer cells to paclitaxel and cisplatin by targeting ABCB1. *J. Ovarian Res.* **2015**, *8*, 80. [CrossRef] [PubMed]

35. Kim, C.W.; Go, R.E.; Hwang, K.A.; Bae, O.N.; Lee, K.; Choi, K.C. Effects of cigarette smoke extracts on apoptosis and oxidative stress in two models of ovarian cancer in vitro. *Toxicol. In Vitro* **2018**, *52*, 161–169. [CrossRef] [PubMed]

36. Langdon, S.P.; Gourley, C.; Gabra, H.; Stanley, B. Endocrine therapy in epithelial ovarian cancer. *Expert Rev. Anticancer Ther.* **2017**, *17*, 109–117. [CrossRef] [PubMed]

37. Bjornstrom, L.; Sjoberg, M. Mechanisms of estrogen receptor signaling: Convergence of genomic and nongenomic actions on target genes. *Mol. Endocrinol.* **2005**, *19*, 833–842. [CrossRef] [PubMed]

38. Heinzelmann-Schwarz, V.; Knipprath Meszaros, A.; Stadlmann, S.; Jacob, F.; Schoetzau, A.; Russell, K.; Friedlander, M.; Singer, G.; Vetter, M. Letrozole may be a valuable maintenance treatment in high-grade serous ovarian cancer patients. *Gynecol. Oncol.* **2018**, *148*, 79–85. [CrossRef] [PubMed]

39. Bhattacharjee, S.; Nandi, S. Synthetic lethality in DNA repair network: A novel avenue in targeted cancer therapy and combination therapeutics. *IUBMB Life* **2017**, *69*, 929–937. [CrossRef]

40. Bhattacharjee, S.; Nandi, S. Choices have consequences: The nexus between DNA repair pathways and genomic instability in cancer. *Clin. Transl. Med.* **2016**, *5*, 45. [CrossRef]

41. Bhattacharjee, S.; Nandi, S. DNA damage response and cancer therapeutics through the lens of the Fanconi Anemia DNA repair pathway. *Cell Commun. Signal.* **2017**, *15*, 41. [CrossRef]

42. Bhattacharjee, S.; Nandi, S. Rare Genetic Diseases with Defects in DNA Repair: Opportunities and Challenges in Orphan Drug Development for Targeted Cancer Therapy. *Cancers* **2018**, *10*, 298. [CrossRef] [PubMed]
43. Scholz, C.; Heublein, S.; Lenhard, M.; Friese, K.; Mayr, D.; Jeschke, U. Glycodelin A is a prognostic marker to predict poor outcome in advanced stage ovarian cancer patients. *BMC Res. Notes* **2012**, *5*, 551. [CrossRef] [PubMed]
44. Nakas, C.T.; Alonzo, T.A.; Yiannoutsos, C.T. Accuracy and cut-off point selection in three-class classification problems using a generalization of the Youden index. *Stat. Med.* **2010**, *29*, 2946–2955. [CrossRef] [PubMed]
45. Youden, W.J. Index for rating diagnostic tests. *Cancer* **1950**, *3*, 32–35. [CrossRef]
46. Perkins, N.J.; Schisterman, E.F. The Inconsistency of "Optimal" Cut-points Using Two ROC Based Criteria. *Am. J. Epidemiol.* **2006**, *163*, 670–675. [CrossRef] [PubMed]
47. Fluss, R.; Faraggi, D.; Reiser, B. Estimation of the Youden Index and its Associated Cutoff Point. *Biom J.* **2005**, *47*, 458–472. [CrossRef] [PubMed]

International Journal of
Molecular Sciences

Article

Ethinylestradiol and Levonorgestrel as Active Agents in Normal Skin, and Pathological Conditions Induced by UVB Exposure: In Vitro and In Ovo Assessments

Dorina Coricovac [1,†], Claudia Farcas [1,†], Cristian Nica [2], Iulia Pinzaru [1,*], Sebastian Simu [1], Dana Stoian [2], Codruta Soica [1], Maria Proks [1], Stefana Avram [1], Dan Navolan [2], Catalin Dumitru [2], Ramona Amina Popovici [3] and Cristina Adriana Dehelean [1]

[1] Faculty of Pharmacy, "Victor Babeş" University of Medicine and Pharmacy, 300041 Timişoara, Romania; dorinacoricovac@umft.ro (D.C.); farcas.claudia@umft.ro (C.F.); simu.sebastian@umft.ro (S.S.); codrutasoica@umft.ro (C.S.); proks.maria@yahoo.ro (M.P.); stefana.avram@umft.ro (S.A.); cadehelean@umft.ro (C.A.D.)

[2] Faculty of Medicine, "Victor Babeş" University of Medicine and Pharmacy, 300041 Timişoara, Romania; nicacristian500@gmail.com (C.N.); stoian.dana@umft.ro (D.S.); navolan@yahoo.com (D.N.); dumcatal@yahoo.com (C.D.)

[3] Faculty of Dentistry, "Victor Babeş" University of Medicine and Pharmacy, 300041 Timişoara, Romania; ramona.popovici@umft.ro

[*] Correspondence: iuliapinzaru@umft.ro; Tel.: +40-725-943-894

[†] These authors contributed equally to this work.

Received: 14 September 2018; Accepted: 10 November 2018; Published: 14 November 2018

Abstract: The link between melanoma development and the use of oral combined contraceptives is not fully elucidated, and the data concerning this issue are scarce and controversial. In the present study, we show that the components of oral contraceptives, ethinylestradiol (EE), levonorgestrel (LNG), and their combination (EE + LNG) ± UVB (ultraviolet B radiation) induced differential effects on healthy (human keratinocytes, fibroblasts, and primary epidermal melanocytes, and murine epidermis cells) and melanoma cells (human—A375 and murine—B164A5), as follows: (i) at low doses (1 µM), the hormones were devoid of significant toxicity on healthy cells, but in melanoma cells, they triggered cell death via apoptosis; (ii) higher doses (10 µM) were associated with cytotoxicity in all cells, the most affected being the melanoma cells; (iii) UVB irradiation proved to be toxic for all types of cells; (iv) UVB irradiation + hormonal stimulation led to a synergistic cytotoxicity in the case of human melanoma cells—A375 and improved viability rates of healthy and B164A5 cells. A weak irritant potential exerted by EE and EE + LNG (10 µM) was assessed by the means of a chick chorioallantoic membrane assay. Further studies are required to elucidate the hormones' cell type-dependent antimelanoma effect and the role played by melanin in this context.

Keywords: ethinylestradiol; levonorgestrel; keratinocytes; fibroblasts; melanocytes; melanoma; ultraviolet radiation

---

1. Introduction

The admission in use of the oral hormonal contraceptives in the 1960s marked a new period in the pregnancy prevention methods [1]. Despite the progress recorded in the field of contraception methods (intrauterine devices, weekly transdermal patches, long-acting hormone-releasing implants, and monthly vaginal rings), combined oral contraceptives continues to be the most preferred form of reversible hormonal birth control [2]. These pills consist of an estrogen and a progestin component [3]. The estrogen component existent in most of the past and current combined oral contraceptives is 17α-ethinylestradiol (EE), a semisynthetic estrogen, obtained in 1938 by substitution of estradiol at

C17 with an ethinyl group, and is described as the most widely used orally bioactive estrogen [2–5]. Levonorgestrel (LNG—13β-ethyl-17α-ethynyl-17β-hydroxy-4-gonen-3-one) is a second-generation synthetic progestogen, a component of oral contraceptives, and included in "category X" of the drugs forbidden in pregnancy [6].

Even after more than five decades of use by an impressive number of women worldwide (hundreds of millions), the safety issues of oral contraceptives still represents a serious matter of concern. The cardiovascular side-effects related to estrogen component were decreased by the gradual decline of ethinylestradiol dosage from 50 to 20 and even 15 µg, whereas the carcinogenic potential of these pills is still debatable [4,7].

Past and recent studies reported that the role of oral contraceptives in the development of cancer (incompletely elucidated) might be considered tissue-type dependent, since after their use, a protective effect was observed by decreasing the risk of endometrial, ovarian, and colorectal cancers, and an increased risk of breast, cervical, and liver cancer was reported [1,7]. In the last few years, growing clinical and experimental evidence regarding the implication of estrogens in skin cancers, and mainly in melanoma development was gathered, but the data on this subject are still scarce and controversial [8–10]. A recent population-based case-control study presented data that support the hypothesis that a long period of use of oral contraceptives (especially with high concentrations of estrogen > 50 mg) is associated with an increased risk of a keratinocyte-derived cancer (squamous cell carcinoma—SCC and basal cell carcinoma—BCC) occurrence [11].

Melanoma is characterized as one of the most immunogenic malignancies, based on its histological, clinical, and genetic heterogeneity that leads to drug resistance to current therapies, reduced tumor regression and survival rates, and converts it into a very demanding challenge [12]. To elucidate the complexity of melanoma growth and progression, novel theories were suggested, consisting of a new approach that describes melanoma as a hormone-related cancer type [9]. This approach might be supported by the following arguments: (i) a gender disparity was observed in melanoma (a reduced incidence, better prognosis and increased survival outcome in female population) [13]; (ii) the estrogen receptors (mainly estrogen receptor β—ERβ) are located in epidermal keratinocytes, dermal fibroblasts, and melanocytes, receptors that mediate key signaling pathways involved in cell proliferation and differentiation, wound healing, skin immune response, and protection against skin photoaging [13–16]; (iii) estrogens play an important role in cell pigmentation activity [17], an impairment of this function leading to the development of melasma (a disorder of melanogenesis characterized by the presence of an increased number of active melanocytes), or even melanoma [18]; (iv) stimulation of ERβ inhibits the proliferation and migration of malignant cells; the loss of ERβ in melanoma and estrogen-related tumors causes a diminished inhibition of malignant melanocytes proliferation, which in turn is stimulated by ERα [19]; (v) progesterone receptors were found in some keratinocytes and in the nuclei of basal cells [20], and controversial results (stimulatory vs. inhibitory effects) were obtained in terms of melanocyte proliferation as a result of progesterone activity [17]. However, the role played by estrogens and progestins in melanoma development is still uncertain; a large study published in 2017 revealed that estrogens alone increase the risk of melanoma, while the estrogen–progestin combined therapy exerted an opposite activity in terms of melanoma development [21]. Another recently published study (2017) showed that the expression of estrogen and progesterone receptors is a sporadic phenomenon in some cases of malignant melanoma [10]. Natale et al. reported that endogenous estrogen and progesterone mutually regulate melanin synthesis through membrane-bound receptors, even in the absence of classical estrogen or progesterone receptors [22].

One of the main risk factors associated with melanoma development is natural or artificial ultraviolet radiation [23] while sex steroid hormones, both endogenous and exogenous are considered as secondary risk factors [21,23]. The harmful activity of ultraviolet radiation on the human body, recognized as a carcinogen agent, may remain inactive for many years until exposure to certain promoters. Hormones and oncogenes are the most eloquent examples of such agents, with estrogens being labeled both as mutagenic agents of the DNA, and as promoters of cell specific alterations [24,25].

An interplay between female sex steroid hormones and UVB irradiation was discussed within the literature, but a direct link between these agents and the development of melanoma has not yet been found. The regular intake of a levonorgestrel–ethinylestradiol combination led to a phototoxic reaction at the skin level, due to a high absorption of UVB and UVA radiation, both for the two substances and for their combination [26], respectively. In vivo estrogen and UVB exposure produced inflammatory mediators in the skin, and thus led to an improper physiological skin response to UV radiation [27]; thus, UVB radiation may induce differential effects of estrogens on the skin [28]. On the contrary, estrogens showed a protective role at the skin level against UVB chronic irradiation, by employing various mechanisms [29].

Another key player in melanoma development is melanogenesis, a metabolic pathway that is specific for both normal and malignant melanocytes, that interferes with melanoma cell behavior and their surrounding environment [30–32]. Melanin is considered to be "a double edge sword" by acting as a protector of melanocytes against UVB deleterious effects and oxidative stress, and on the other hand, a deregulated melanogenesis leads to an increased melanoma resistance to therapy (melanogenesis intermediates exert mutagenic, genotoxic and immunosuppressive properties, and induce hypoxia by upregulating HIF-1α expression) [31,32]. Moreover, it was proven that amelanotic melanomas exhibit a higher susceptibility to radiotherapy as compared to melanotic melanomas, and the overall survival of the patients with amelanotic melanomas is increased [33,34].

Taken together, with all of the information stated above, we could conclude that the current data regarding the association of oral combined hormonal therapy, UV radiation, and skin malignancies is still poor. In this study, we focused on the evaluation of the cytotoxic profile of ethinylestradiol (EE), levonorgestrel (LNG), and their association (EE + LNG), with and without UVB irradiation, on healthy cell lines (human keratinocytes, fibroblasts, and primary epidermal melanocytes, and mouse epidermis cells) and tumor cell lines (human and murine melanoma) by in vitro (viability, migration and proliferation) and in vivo (HET-CAM) techniques.

2. Results

2.1. Ethinylestradiol and Levonorgestrel ± UVB Irradiation Induced Differential Effects on Healthy Cell and Tumor Cell Viability

To assess the effect induced by test compounds (EE, LNG and EE + LNG) on healthy human and murine skin (HaCaT, 1BR3, HEMa and JB6 Cl 41-5a) cells, and melanoma (A375 and B164A5) cell viability in the presence/absence of UVB irradiation, we performed the Alamar blue assay. Irradiation of HaCaT, 1BR3, HEMa and JB6 Cl 41-5a cells with UVB (40 mJ/cm^2) resulted in a significant reduction of cells viability (66.30% viable HaCaT, 74.75% viable 1BR3, 58.25% viable HEMa, and 60.85% viable JB6 Cl 41-5a, respectively) as compared to control cells (unirradiated cells) (Figures 1 and 2). Stimulation of healthy cells with EE (1 and 10 μM) for 24 h led to the following results: (i) HaCaT cells—a slight decrease of viability in a dose-dependent manner (92.90% at 1 μM and 82.01% at 10 μM), (ii) 1BR3 cells—88.04% viable cells at 10 μM, (iii) HEMa cells—82.25% viable cells at 10 μM, and (iv) JB6 Cl 41-5a cells—the viability was not affected as compared to control cells (unstimulated cells) (Figures 1 and 2). Levonorgestrel had no influence on HaCaT, 1BR3, and JB6 Cl 41-5a cell viability after 24 h stimulation at the lowest concentration tested—1 μM, whereas at 10 μM it was recorded a decrease <10% in the case of HaCaT cells and <5% in the case of 1BR3 and JB6 Cl 41-5a. In the case of HEMa cells, the effect of levonorgestrel was somehow reversed as compared to the other healthy cells: the lowest concentration—1 μM decreased cells viability (81.69%), whereas at 10 μM, no toxicity was observed. A combination of EE + LNG induced a decline of JB6 Cl 41-5a viability (around 10%) at both tested concentrations (1 and 10 μM), with HaCaT, 1BR3, and HEMa cells being affected only at the highest concentrations (<10% decrease for HaCaT and 1BR3 and <5% decrease for HEMa) (Figures 1 and 2).

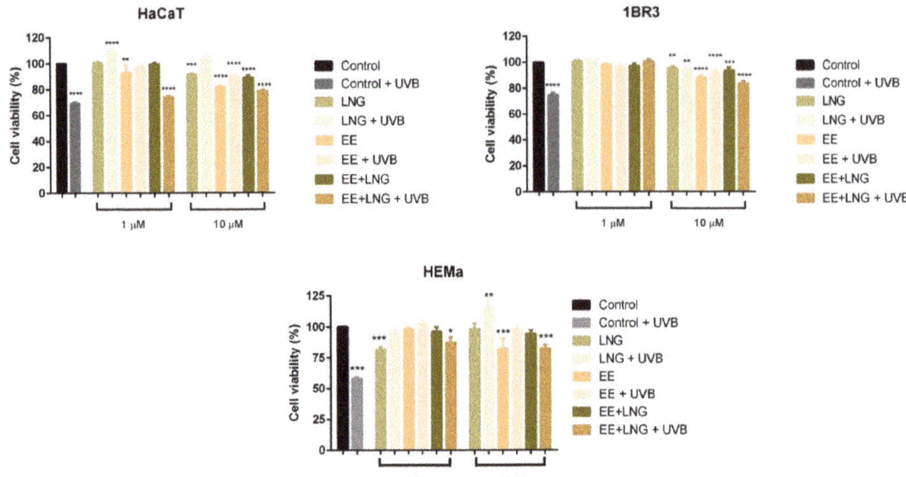

Figure 1. The effect of test compounds (1 and 10 μM) ± UVB irradiation on HaCaT—human keratinocytes, 1BR3—human skin fibroblasts and HEMa—primary human epidermal melanocytes viability at 24 h post-stimulation. The results are expressed as cell viability percentage (%) normalized to control cells. The data represent the mean values ± SD of three independent experiments. One-way analysis of variance (ANOVA) analysis was applied to determine the statistical differences followed by Tukey's multiple comparisons test (* $p < 0.05$; ** $p < 0.01$; *** $p < 0.001$, **** $p < 0.0001$). EE: ethinylestradiol; LNG: levonorgestrel.

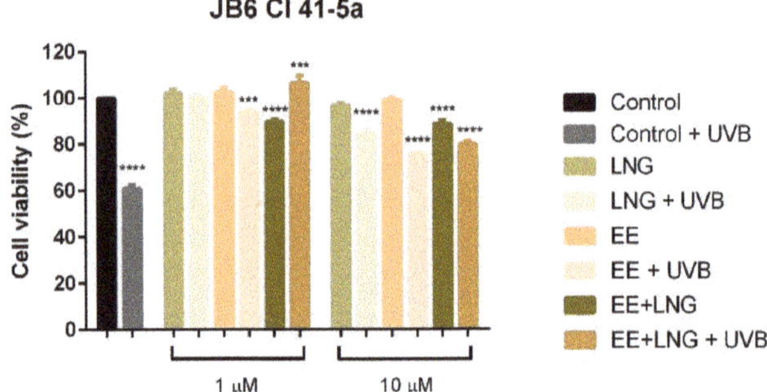

Figure 2. The effect of test compounds (1 and 10 μM) ± UVB irradiation on JB6 Cl 41-5a cell viability at 24 h post-stimulation. The results are expressed as cell viability percentage (%) normalized to control cells. The data represent the mean values ± SD of three independent experiments. One-way ANOVA analysis was applied to determine the statistical differences followed by Tukey's multiple comparisons test (*** $p < 0.001$, **** $p < 0.0001$).

The lowest viability rates were observed in the groups of cells that were irradiated with UVB and stimulated with the combination of hormones—EE + LNG (at 10 μM); still, these viability percentages were higher as compared to the ones recorded for the cells that were only UVB-exposed (HaCaT: 78.55% vs. 69.30%; 1BR3: 83.31% vs. 74.75%, HEMa: 82.46% vs. 58.25%, and JB6 Cl 41-5a: 79.83% vs. 60.85%), what might indicate a recovery effect induced by EE + LNG stimulation after UVB noxious effects on healthy skin cells (see Figures 1 and 2).

Similar experimental conditions to the ones described for healthy cells were applied for A375 and B164A5 melanoma cells in order to evaluate the effects induced by test compounds (1 and 10 μM) ± UVB irradiation on cells viability in a 24 h frame.

Results showed that UVB irradiation of human and murine melanoma cells determined a significant decrease of cell viability (around 75%) as compared to control cells (unirradiated cells) (Figure 3). Both EE and LNG induced a dose-dependent decline of A375 and B164A5 cell viability, but the lowest viability percentage was calculated for the EE + LNG at the highest concentration used—10 μM (56% for A375 and 47.23% for B164A5). Exposure to UVB radiation followed by stimulation with EE, LNG, or EE + LNG led to a significant dose-dependent decrease of A375 cell viability percentage, decrease that was considerably stronger as compared to the effects induced by each test compound/UVB alone, what might lead to the conclusion that the used agents had a synergistic cytotoxic effect on A375 cells (EE vs. EE + UVB: 66.54% vs. 58.72%; LNG vs. LNG + UVB: 69.78% vs. 67.59%; EE + LNG vs. EE + LNG + UVB: 56% vs. 49.69%). In the case of B164A5 cells, UVB irradiation followed by stimulation with test compounds produced an inverse effect as compared to A375; namely, an increase of the cells' viability as compared with the values obtained for each test compound (EE vs. EE + UVB: 56.84% vs. 74.46%; LNG vs. LNG + UVB: 59.27% vs. 78.06%; EE + LNG vs. EE + LNG + UVB: 47.23% vs. 80.59%) (Figure 3). A similar effect as the one described for B164A5 was observed in the case of pigmented human melanoma cells—RPMI-7951 (see Figure S1).

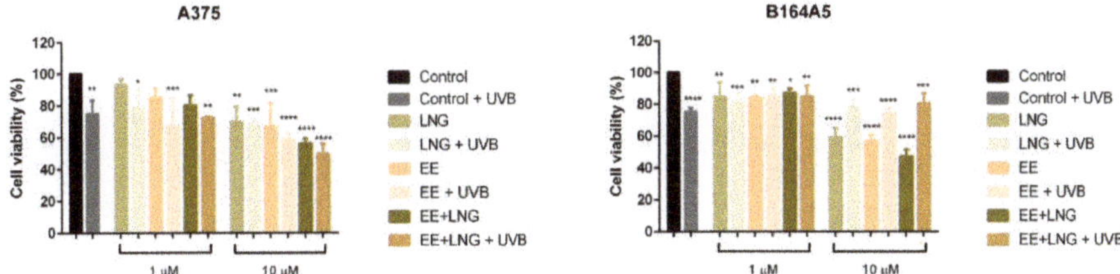

Figure 3. The effect of test compounds (1 and 10 μM) ± UVB irradiation on A375—human melanoma and B164A5—murine melanoma cell viability at 24 h post-stimulation. The results are expressed as a cell viability percentage (%) normalized to control cells. The data represent the mean values ± SD of three independent experiments. One-way ANOVA analysis was applied to determine the statistical differences followed by Tukey's multiple comparisons test (* $p < 0.05$; ** $p < 0.01$; *** $p < 0.001$, **** $p < 0.0001$).

2.2. Ethinylestradiol and Levonorgestrel ± UVB Irradiation Triggered Apoptosis in A375 and B164A5 Melanoma Cells

Based on the results described above, according to which the test compounds (EE, LNG, EE + LNG) ± UVB significantly decreased the viability of human and murine melanoma cells, it was verified if the cells death was achieved via apoptosis; the analysis was performed using an annexin V/PI (propidium iodide) apoptosis detection kit. The cells were stimulated for 24 h with EE, LNG and EE + LNG (1 and 10 μM) ± UVB irradiation.

A dose-dependent apoptotic activity was noticed in the case of both cell lines. As compared to control cells (unstimulated cells), the strongest apoptotic effect on non UVB-irradiated A375 human melanoma cells was induced by EE and EE + LNG at the highest concentrations tested—10 μM); the percentage of early apoptotic cells was 51.78% for EE and 51.15% for EE + LNG (Figure 4), data that confirm the results obtained for viability assessment. At the same concentration, LNG alone exerted a lower pro-apoptotic activity. UVB exposure of A375 cells, followed by addition of 1 μM of test compounds led to a significantly increased percentage of early apoptotic cells as follows: 22.62% for LNG; 31% for EE and 27% for EE + LNG. UVB irradiation combined with the highest concentration—10 μM of test compounds triggered percentages of the early apoptotic population similar to the ones recorded for the test compounds in non UVB-exposed cell population (Figure 4).

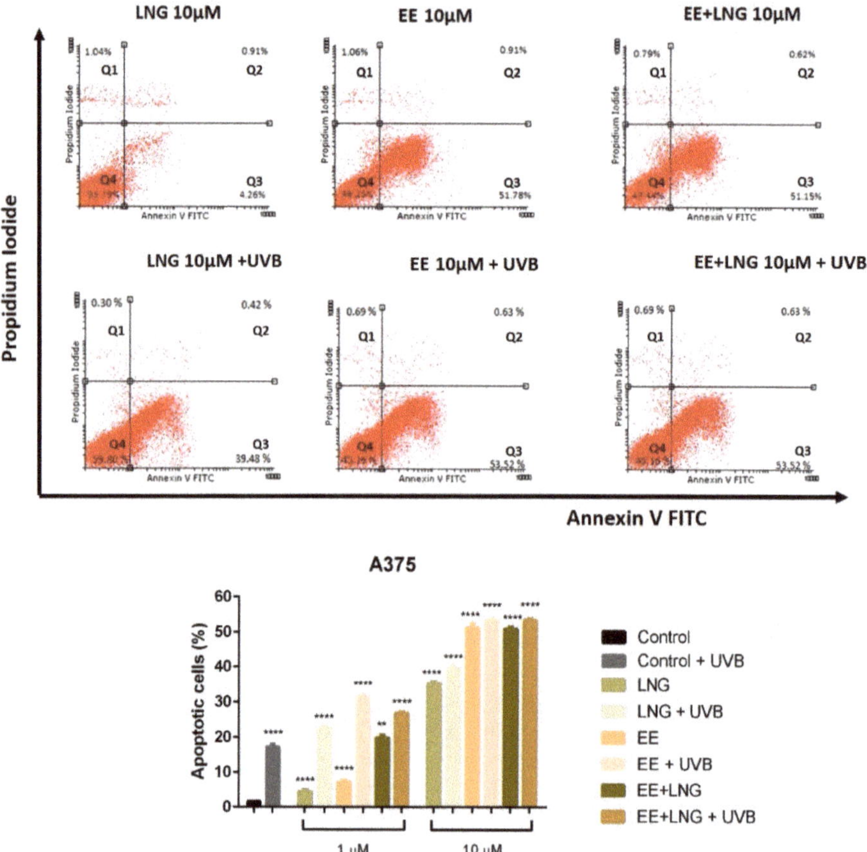

Figure 4. Representative dot plots of the apoptotic events induced by test compounds (EE, LNG, EE + LNG—10 µM) ± UVB irradiation in A375 human melanoma cells after a 24 h stimulation. The cells status was analyzed by a FACS technique where: Q4—viable cells, Q3—early apoptotic cells, Q2—late apoptotic cells and Q1—necrotic cells. The graph represents the percentage of early apoptotic A375 cells. The results are expressed as apoptotic cell percentage (%) normalized to control cells. The data represent the mean values ± SD of three independent experiments. One-way ANOVA analysis was applied to determine the statistical differences, followed by Tukey's multiple comparisons test (** $p < 0.01$; **** $p < 0.0001$).

In Figure 5 was depicted the effect of the test compounds ± UVB irradiation on B164A5 murine melanoma cells apoptotic process; the highest concentration tested—10 µM induced a drastic decrease of cell viability and caused the most significant pro-apoptotic effect with a maximum of 72.83% for EE + LNG. After UVB exposure and 1 µM of test compounds, one can notice the absence of the pro-apoptotic process and the subsequently increased cell viability. At 10 µM, B164A5 UVB-irradiated murine melanoma cells showed a slight apoptosis induction, with the highest pro-apoptotic level noted for EE (21.44%) (Figure 5).

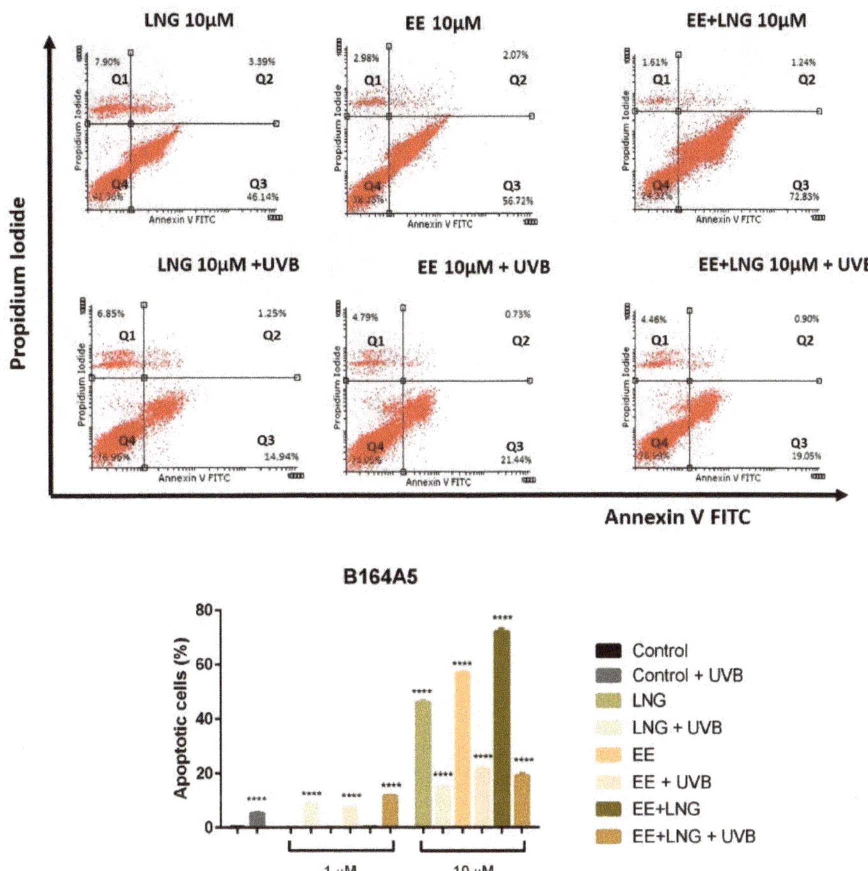

Figure 5. Representative dot plots of the apoptotic events induced by test compounds (EE, LNG, EE + LNG—10 µM) ± UVB irradiation in B164A5 murine melanoma cells after a 24 h stimulation. The cells status was analyzed by a FACS technique where: Q4—viable cells, Q3—early apoptotic cells, Q2—late apoptotic cells and Q1—necrotic cells. The graph represents the percentage of early apoptotic B164A5 cells. The results are expressed as apoptotic cell percentage (%) normalized to control cells. The data represent the mean values ± SD of three independent experiments. A one-way ANOVA analysis was applied to determine the statistical differences followed by Tukey's multiple comparisons test (**** $p < 0.0001$).

2.3. Ethinylestradiol (EE) and Levonorgestrel (LNG) ± UVB Irradiation Determined Changes in Cells Morphology

Immortalized human keratinocytes—HaCaT showed no significant morphological changes after stimulation with EE, LNG, and EE + LNG (1 µM). Their shape remained well defined, elongated, and the cells were attached to the culture plate. In contrast, after UVB irradiation, HaCaT cells drastically changed their morphological aspect, becoming round and some of them floating; the most affected cells seemed to be the ones stimulated with EE + LNG. Cell shrinkage was also noticed, and could be considered a sign of early apoptosis, results that are consistent with the data described in the apoptosis assessment section. At 24 h post-exposure to UVB, HaCaT cells stimulated with EE and LNG looked like they began to partially recover, results that confirm the cell viability data (Figure 6).

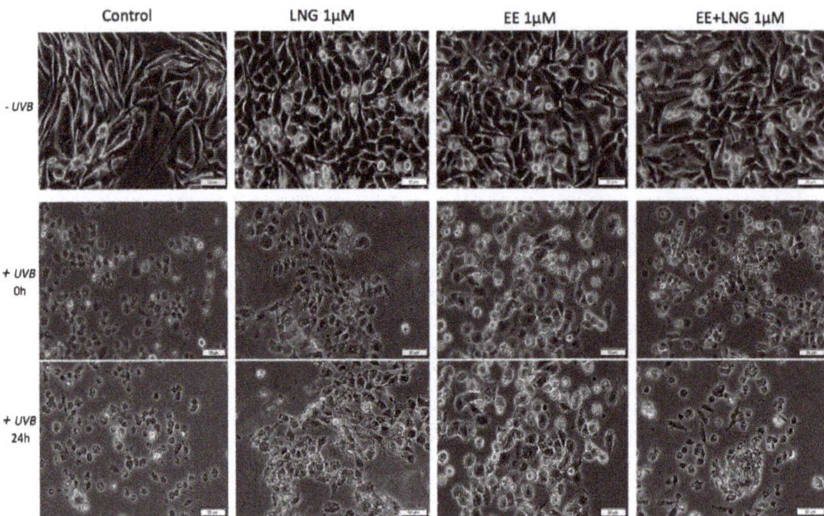

Figure 6. In vitro morphological aspects of human keratinocytes—HaCaT cells, stimulated with levonorgestrel (LNG), ethinylestradiol (EE), and an ethinylestradiol/levonorgestrel combination (EE + LNG), respectively, at a concentration of 1 µM ± UVB irradiation. Scale bars represent 50 µM.

In the case of human skin fibroblasts—1BR3, the results were similar as for HaCaT cells, with no changes in cells shape following stimulation with EE, LNG, and EE + LNG were noticed; the cells morphology preserving the same needle-like shape and the same confluence as the control cells. Control cells exposed to UVB showed various degrees of cell shrinkage; but after stimulation with EE, LNG, and EE + LNG, respectively, cells began to regain their initial morphological aspect with bright and compact cell margins; however, the colonial morphology was not entirely recovered after 24 h but recovery signs in cells shape were detected (Figure 7).

Stimulation of primary human epidermal melanocytes—HEMa with EE and LNG (1 µM) had no effects on cells morphology, the cells were adherent to the culture plate and presented a needle-like/dendritic-like shape similar to control cells. EE + LNG induced a slight modification of HEMa cells morphology. UVB irradiation influenced the melanocytes' shape and their confluence, and the association with EE + LNG seemed to be the most noxious. At 24 h post-exposure to UVB, HEMa cells stimulated with EE and LNG partially gained their initial form, results that confirm the cell viability data (Figure 8).

Murine epidermis JB6 Cl 41-5a cells showed a good confluence in the absence of UVB radiation and the test compounds did not perturb the shape of the cells; whereas after UVB exposure, the cells stimulated with test compounds seemed to be protected by UVB deleterious effects, and only minor changes were observed in the group stimulated with EE. The control cells exposed to UVB were most affected, displaying a low level of confluence and major changes of their morphological aspects, characteristics that were partially recovered after 24 h post-irradiation (Figure 9).

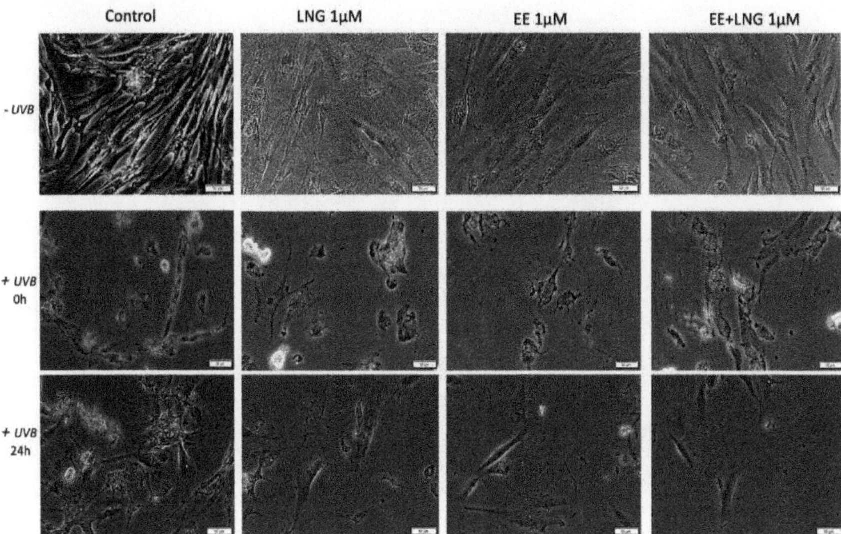

Figure 7. In vitro morphological aspect of human fibroblasts—1BR3 cells, stimulated with levonorgestrel (LNG), ethinylestradiol (EE), and an ethinylestradiol/levonorgestrel combination (EE + LNG), respectively, at a concentration of 1 µM ± UVB irradiation. Scale bars represent 50 µM.

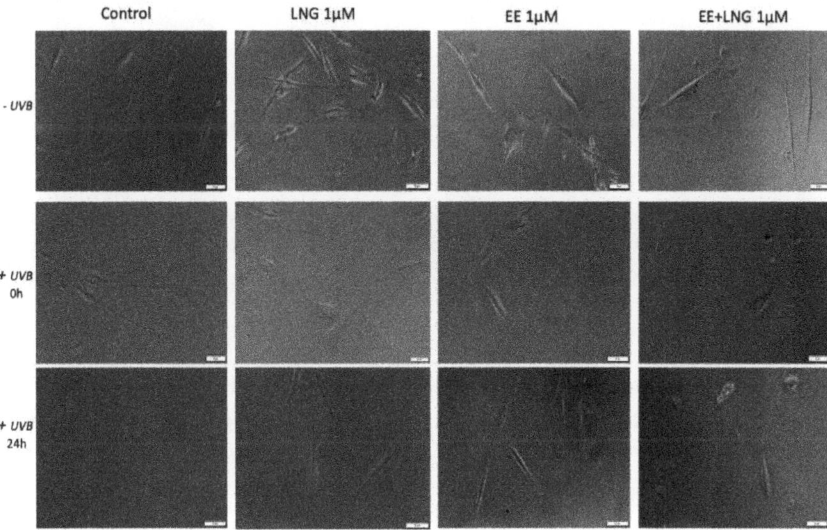

Figure 8. In vitro morphological aspects of primary human epidermal melanocytes—HEMa cells, stimulated with levonorgestrel (LNG), ethinylestradiol (EE), and ethinylestradiol/levonorgestrel combination (EE + LNG), respectively, at a concentration of 1 µM ± UVB irradiation. Scale bars represent 50 µM.

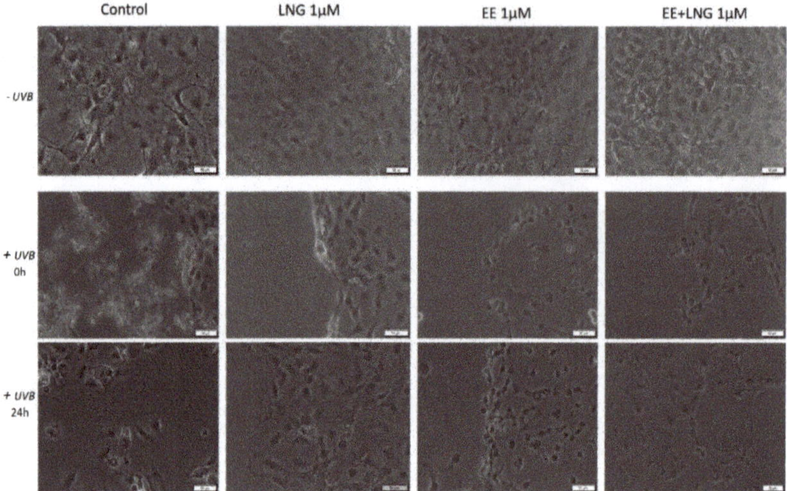

Figure 9. In vitro morphological aspect of mice epidermis—JB6 Cl 41-5a cells, stimulated with levonorgestrel (LNG), ethinylestradiol (EE), and an ethinylestradiol/levonorgestrel combination (EE + LNG), respectively, at a concentration of 1 µM ± UVB irradiation. Scale bars represent 50 µM.

Taking into consideration the pro-apoptotic effect of the test compounds on human and murine melanoma cells, the impact of these compounds on melanoma cells morphology was monitored by light microscopy. In the case of A375, the control cells (unstimulated and unirradiated cells) displayed a normal epithelial morphology, with spindle and cobblestone shapes, strongly bounded, adherent to the culture plate, and highly confluent after 24 h. A decrease of A375 control cells confluence was recorded after UVB radiation and some detached and floating cells were noticed. The EE and LNG stimulation of cells exposed to UVB led to some changes in cells' shape (Figure 10), mainly after EE + LNG treatment; the cells became round and began to detach, indicating the process of apoptosis, the results being in agreement with the reported cell viability data. In the case of pigmented human melanoma cells—RPMI-7951, the test compounds had no impact on cells morphology, but after UVB irradiation, significant changes were observed in all groups (round floating cells), effects that were almost completely reversed after 24 h and test compound stimulation (see Figure S2).

In the case of murine melanoma cells—B164A5, the cells exposed to UVB irradiation seemed to be the most greatly affected in terms of cell morphology, showing a round shape with dendrites and shrinkage. Changes in B164A5 melanoma cells shape were also observed after stimulation with test compounds, in non-UVB irradiated cells.

On the other hand, B164A5 cells exposed to UVB followed by stimulation with test compounds revealed a confluence increment and minor changes in cells morphology, as shown in Figure 11.

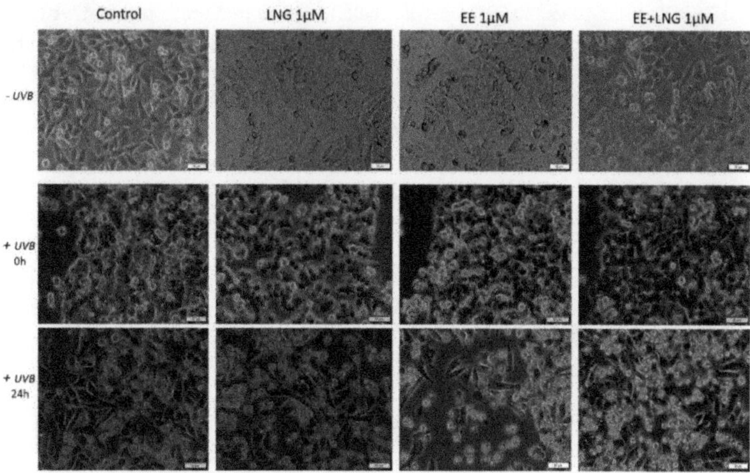

Figure 10. In vitro morphological aspect of human melanoma—A375 cells, stimulated with levonorgestrel (LNG), ethinylestradiol (EE), and an ethinylestradiol/levonorgestrel combination (EE + LNG), respectively, at a concentration of 1 μM ± UVB irradiation. Scale bars represent 50 μM.

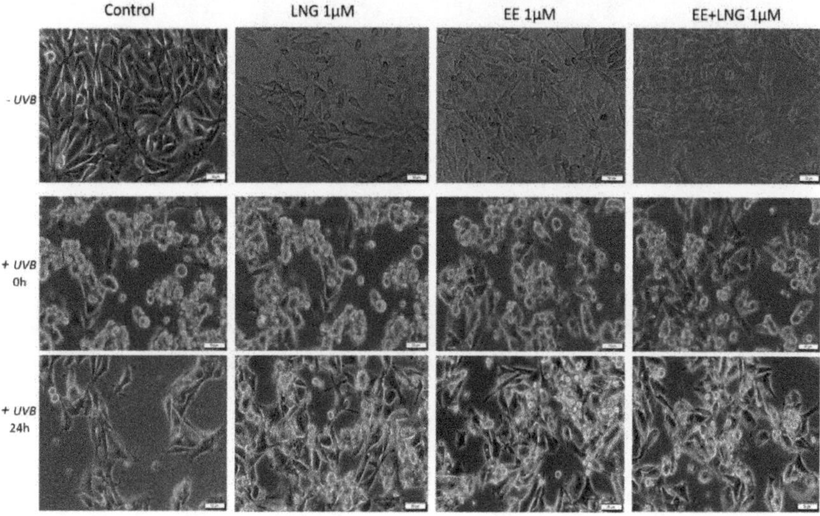

Figure 11. In vitro morphological aspect of murine melanoma—B164A5 cells, stimulated with levonorgestrel (LNG), ethinylestradiol (EE), and an ethinylestradiol/levonorgestrel combination (EE + LNG), respectively, at a concentration of 1 μM ± UVB irradiation. Scale bars represent 50 μM.

2.4. The impact of Ethinylestradiol (EE) and Levonorgestrel (LNG) on Healthy and Tumor Cells Migration and Proliferation

Figure 12 displays the migratory activity of the healthy cell lines in the presence of EE, LNG, and EE + LNG. Since at the highest concentration used—10 μM, a cytotoxic and pro-apoptotic effect was observed, and the concentration selected for this assay was 1 μM. LNG stimulation did not interfere with the migration of human and murine healthy skin cells, the wound widths at 24 h being similar to the ones measured for control cells (Figure 12). After EE stimulation, a stimulatory trend in all cell lines could be mentioned as compared to control cells; however, the most significant stimulation was seen with 1BR3 cells (52.37% vs. 40.09% on 1BR3 cells), results that are consistent with cell viability data. The combination of the two hormones—EE + LNG induced an inhibitory effect on HaCaT cells migration, showing a wound closure rate of 58.18%, whereas in the case of 1BR3 and JB6 Cl 41-5a, the effect was a stimulatory one (Figure 12). The very low wound healing rate (40.08%) of 1BR3 control cells was due to their low proliferation ability in specific culture conditions per day. A stimulatory effect on HEMa cells migration was observed after EE and LNG stimulation (the gap was almost covered—mainly after EE) as compared with control cells. Moreover, the combination EE + LNG also augmented the migratory capacity of HEMa cells (Figure 12).

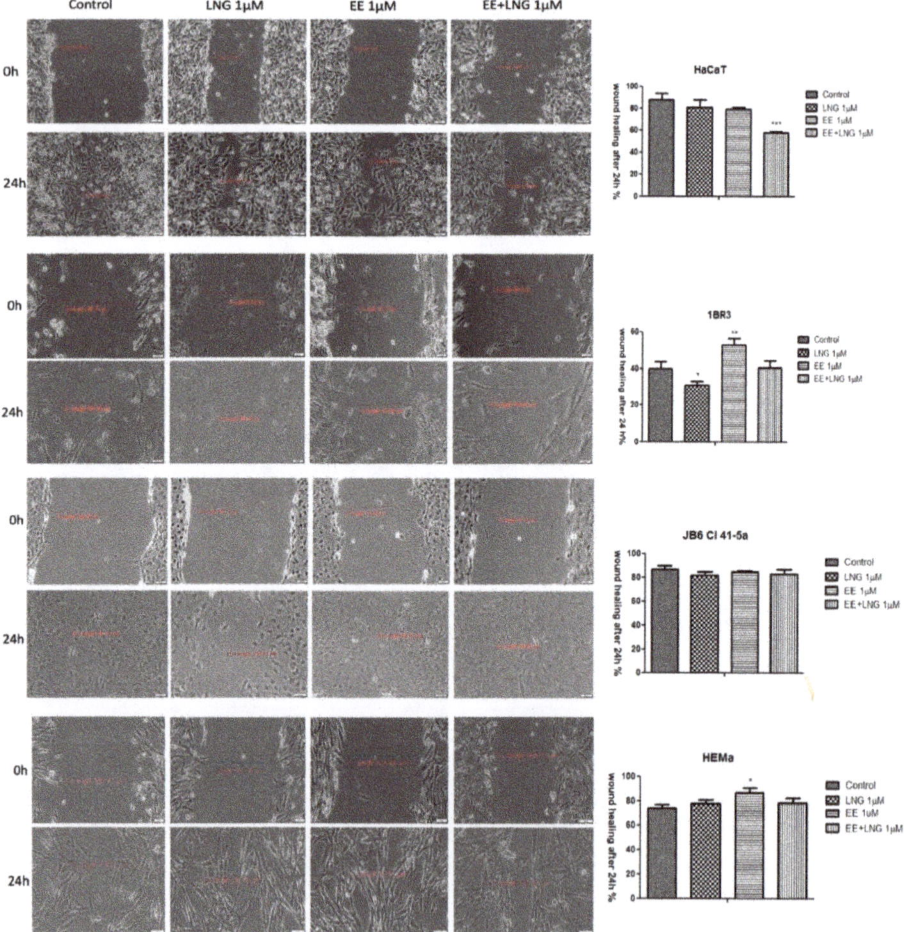

Figure 12. The impact of test compounds (LNG, EE, and EE + LNG—1 μM) on the migratory capacity of healthy skin cell lines (HaCaT—human immortalized keratinocytes, 1BR3—human fibroblasts, JB6Cl415a—mice epidermis, HEMa—primary human epidermal keratinocytes). Wound closure was recorded by bright field microscopy initially—0 h and after 24 h, respectively. Scale bars represent 50 μm. The bar graphs are expressed as percentage of wound closure after 24 h compared to the initial surface. The data represent the mean values ± SD of three independent experiments. One-way ANOVA analysis was applied to determine the statistical differences followed by Tukey post-test (* $p < 0.05$; ** $p < 0.01$; *** $p < 0.001$ vs. control—no stimulation).

The in vitro wound healing assay revealed that the melanoma cells' (A375—human melanoma, B164A5—murine melanoma) migratory capacity was not inhibited by EE and EE + LNG stimulation (1 µM), moreover, a stimulatory effect could be mentioned; still, the fact that the wound was also covered with some detached cells must be taken into account (Figure 13). For EE, the wound closure rate was 82.81% on human melanoma cells and 85.29% on the murine melanoma cell line. In contrast, the same concentration of LNG (1 M) showed a wound healing rate of only 63.98% in the case of human melanoma cells, and 53.94% in the case of the murine melanoma cell line. Similar results were obtained for human pigmented melanoma cells—RPMI-7951 (see Figure S3).

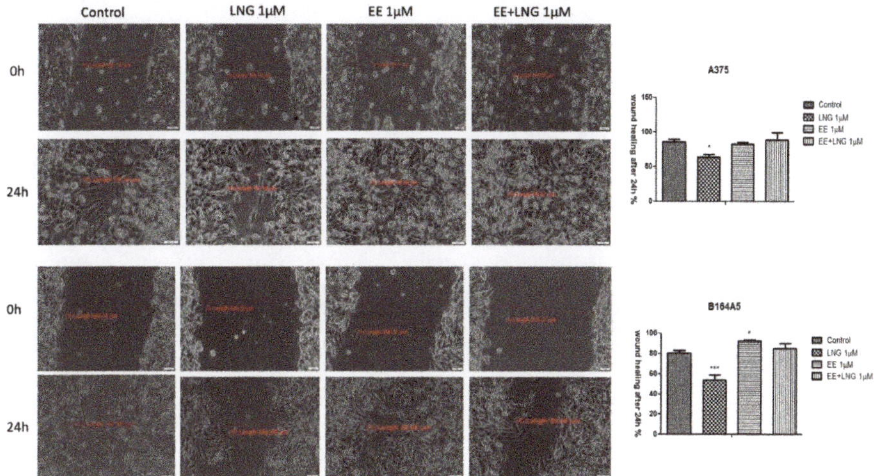

Figure 13. The impact of test compounds (LNG, EE and EE + LNG—1 µM) on migratory capacity of melanoma cell lines (A375—human melanoma cells and B164A5—murine melanoma cells). Wound closure was recorded by bright field microscopy initially and after 24 h, respectively. Scale bars represent 50 µm. The bar graphs are expressed as percentage of wound closure after 24 h compared to the initial surface. The data represent the mean values ± SD of three independent experiments. One-way ANOVA analysis was applied to determine the statistical differences followed by Tukey post-test (* $p < 0.05$; *** $p < 0.001$ vs. control—no stimulation).

2.5. Irritant Potential Assessment of Ethinylestradiol and Levonorgestrel by the Means of a HET-CAM Assay

The potential toxicity of the test compounds (EE, LNG and EE + LNG) was also assessed in vivo, using the in ovo chick chorioallantoic membrane as a biological environment. The protocol allows the evaluation of the irritant potential of the hormone solutions after topical application. The reaction induced by the tested compounds (Table 1) can be classified according to Luepke, as follows: non-irritant (0–0.9), weak irritant (1–4.9), moderate irritant (5–8.9/9.9), and strong irritant (8.9/9.9–21) [35].

The effects induced by the test compounds, along with the positive (SDS—sodium dodecyl sulfate) and negative (PBS—phosphate saline buffer) controls were registered as photographs representing the upper surface of the chorioallantoic membranes before and after 5 min of contact with the solutions. Prior to the determination of the irritation score, the results recorded for irritation severity were considered. SDS induced major vascular damage on the chorioallantoic membrane. All three endpoints: hemorrhage, coagulation, and lysis, were reported only for SDS.

Table 1. The irritant potential of tested hormones: EE, LNG, EE + LNG.

Test Compound and Controls	Irritation Score (Mean)	Irritation Severity (Mean)	Classification of the Effect
PBS Negative control	0 ± 0	0 ± 0	Non-irritant
SDS Positive control	15.07 ± 1.08	2.67 ± 0.58	Strong irritant
DMSO 1% Solvent Control	0 ± 0	0 ± 0	Non-irritant
EE 1 μM	0 ± 0	0 ± 0	Non-irritant
EE 10 μM	2.79 ± 0.55	1.33 ± 0.58	Weak irritant
LNG 1 μM	0 ± 0	0 ± 0	Non-irritant
LNG 10 μM	0 ± 0	0 ± 0	Non-irritant
EE + LNG 1 μM	0.63 ± 0.3	0.83 ± 0.29	Non-irritant
EE + LNG 10 μM	1.23 ± 0.3	1 ± 0	Weak irritant

None of the three endpoints were registered for PBS, DMSO 1%, LNG (1 and 10 μM), and the lowest concentration of EE (1 μM). EE (10 μM) showed late and limited signs of hemorrhage or coagulation, and early, though limited signs of vasodilatation. EE + LNG (1 and 10 μM) application induced slight and limited coagulation, in a dose-dependent manner. The highest mean irritation score was recorded for the positive control, SDS, IS = 14.05. Negative and solvent controls were non-irritant. Among the tested hormones, LNG indicated no sign of irritancy even at the highest concentration tested—10 μM. EE induced a weak irritant effect at the highest concentration (Table 1, Figure 14).

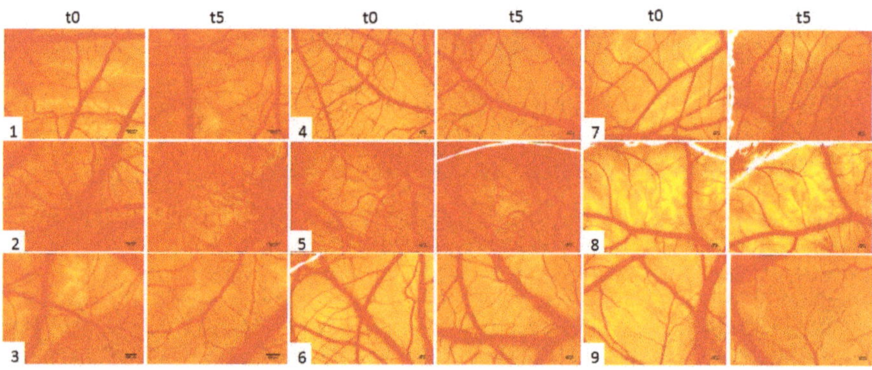

Figure 14. Irritant potential assessment of test compounds using HET-CAM assay: a) stereomicroscope images of the CAMs inoculated with control and test compounds (1—PBS, 2—SDS, 3—DMSO, 4—EE 1 μM, 5—LNG 1 μM, 6—EE + LNG 1 μM, 7—EE 10 μM, 8—LNG 10 μM, 9—EE + LNG 10 μM)—before the application (t0) and after 5 min of contact with the compounds (t5). Scale bars represent 500 μm.

SDS induced major vascular damage on the chorioallantoic membrane; after the application of 500 μL solution, a large area was affected by early micro-hemorrhages, coagulation, and later vessel lysis. The death of the specimen was registered within 60 min. For the samples that were non-irritant on the CAM, we registered a viability of more than 24 h. For the samples that induced a weak irritant effect the death was registered within the first 24 h.

All the tested samples induced no damage or merely slight damages on the CAM vascular plexus. LNG was assessed as non-irritant in both concentrations, EE as non-irritant at 1 μM and a weak irritant at 10 μM. Very similar to EE, the combination EE + LNG was considered non-irritant at 1 μM and a weak irritant (however weaker than EE alone) at 10 μM.

3. Discussion

Oral contraceptives have been suspected for a long time to co-participate in some pathways of developing malignant melanoma, but there was no statistical evidence for either exogenous or endogenous hormones clearly increasing the risk of melanoma [36–38]. The current scientific data are debatable due to studies that confirm the association between sex hormones and melanoma [11,39,40], while others state the opposite [41,42]. According to Nurses' Health Study, the risk to developing melanoma is two times higher among women that have used oral contraception for 10 years or more [43]. It was also reported that the use of progesterone alone actuated the growth of melanoma micro-metastases [44]. Another study revealed that only low doses of progesterone (up to 1 μM), similar to the ones used in therapy, are able to stimulate melanoma cell proliferation, while higher doses not only lack such effect but even induce cell cycle arrest and apoptosis [40].

The myriad of biological and environmental factors that are suspected to interfere in melanoma development leaves a wide-open window for hypotheses, and recent studies investigate estrogen-mediated signaling in melanoma (an impairment of estrogen signaling triggers cancer initiation, promotion and progression) [9], by assessing the role of ERα gene promoter methylation or the expression of G protein-coupled ER [10]. Some reports endorse the existence of a direct relationship between skin diseases, endocrinology, and psychological stress [45]. Moreover, a strong interdependence was reported between the stratum corneum integrity, hormonal levels, and UV susceptibility in terms of minimal erythemal doses, therefore suggesting a significant relevance for all these factors in skin pathophysiology [46].

Most of the experimental studies conducted to verify the role of estrogens and progestins in skin biology/pathology employed as test agents: 17β-estradiol (E2) [47,48] and progesterone (endogenous hormones) [40,49], and data regarding the effects of synthetic hormones present in the composition of oral contraceptives, are rather scanty.

All these converging elements determined the implementation of the present study, which was designed to characterize the in vitro and in ovo toxicological profile of the most frequently used synthetic hormones (ethinylestradiol and levonorgestrel) in oral combined contraceptives by applying two different settings: (1) stimulation with EE, LNG, and EE + LNG of healthy skin cells (human keratinocytes, fibroblasts, primary melanocytes, and murine epidermis cells), melanoma cells (human and murine) and chorioallantoic chick membrane; and (2) healthy and tumor cell UVB irradiation (a well-known initiator and promoter of skin cancer), followed by hormone stimulation for 24 h.

The healthy cell lines used in the experiment (HaCaT—immortalized human keratinocytes, 1BR3—human dermal fibroblasts and JB6 Cl 41-5a—mice epidermis cells) were selected based on the following considerations: (i) the presence of estrogen receptors (ERβ) in epidermal keratinocytes and dermal fibroblasts, the main cellular processes at this level being mediated by estrogens; (ii) estrogens exert a stimulatory effect on melanocytes (estrogen-responsive cells) [10]; (iii) keratinocytes and fibroblasts interact in a synergistically manner to maintain a functional epidermis by promoting repair and regeneration post-acute UVB irradiation [50]; (iv) keratinocytes promote UV-induced melanogenesis (tanning) by releasing several pro-pigmenting paracrine growth factors (αMSH, ET-1, and SCF); (v) dermal fibroblasts are involved in the regulation of constitutive pigmentation and in the development of pigmentary disorders [51].

A primary human epidermal melanocytes cell line—HEMa, was also included in the study, taking in consideration the fact that melanogenesis and melanoma development are strongly interrelated [30–34]. In addition, there is evidence that estrogen and progesterone regulate melanin synthesis [22].

Stimulation of healthy cells with EE, LNG, and EE + LNG led to cell type-dependent results, as follows: HaCaT cells were sensitive to EE in a dose-dependent manner, while LNG and EE + LNG affected cells viability only at the highest concentration (10 μM) (see Figure 1); in the case of 1BR3 cells, the tested hormones reduced cells viability only at the highest concentration; HEMa cells were sensitive to LNG (1 μM) and EE (10 μM), whereas EE + LNG did not decrease melanocyte viability (see Figure 1), and the JB6 Cl 41-5a cells proved to be sensitive only after LNG and EE + LNG stimulation (10 μM) (see Figure 2). Altogether, our results indicate that the lowest concentration (1 μM) of the tested hormones and their combination could be considered without significant toxicity on healthy cells viability, but an increased concentration could affect this status (approximately 75–90% viable cells at 10 µM). A decreased percentage of viable HaCaT cells was also reported after stimulation with high concentrations of EE [52], data that are consistent with our results. The endogenous estrogenic hormone, 17β-estradiol stimulated the proliferation of human normal keratinocytes by augmenting the proportion of cells in S phase of the cell-cycle [53]. The different cellular response observed after EE and 17β-estradiol stimulation could be explained by the fact that EE predominantly acts on ERα whereas 17β-estradiol is equally active on both ERα and ERβ [5].

Several studies reported beneficial and protective roles of estrogens in skin biology (augmented wound healing, protection against photoaging, increased epidermal thickness, ameliorated inflammatory pathologies) initiated via ERα (particularly detected in sebocytes) and ERβ (highly expressed in various skin cell types) [13,47]. It was also stated that estrogens intervene in cell migration and the protection of cell integrity by controlling cell morphology and inducing the cytoskeleton reorganization of different normal and tumor cell types: human dermal fibroblasts (actin cytoskeleton reorganization, restoration of cell shape cultivated in desteroidated medium, and protection on cells adhesive strength), glial cells, neurons, endothelial cells, osteoblasts, and carcinoma cells [47]. A stimulatory effect on healthy cells (HaCaT, 1BR3 and JB6 Cl 41-5a) migration was observed after stimulation with EE (1 µM); however, EE + LNG induced a slight inhibition of HaCaT cells migratory capacity, and LNG did not influence this process (Figure 12). No morphological changes of healthy cells were noticed after hormones stimulation (Figures 6–9).

Concerning the behavior of sex hormones on healthy cells, a recent study demonstrated that a continuous exposure of melanocytes to estrogen led to an increase in melanin production, while progesterone had inverse effects. Moreover, estrogen-treated melanocytes produced a high amount of melanin for 50 days after hormone removal, but in the case of progesterone the cells returned to their baseline level of melanin immediately. In addition, in the melanocytes treated with estrogen, stimulation with progesterone reversed estrogen effects [48]. Similar results were obtained by Wiedeman et al. [52] data that are in agreement with our results. Poletini and co-workers proved in an elegant study that normal and malignant melanocytes respond different to estradiol stimulation [54].

The second setting proposed in this study that involves UVB irradiation determined significant changes in terms of healthy cell viability and morphology. UVB irradiation (40 mJ/cm^2) reduced significantly the percentage of HaCaT, 1BR3, HEMa, and JB6 Cl 41-5a viable cells, the highest toxicity being recorded for HEMa—58.25% and JB6 Cl 41-5a cells—60% viable cells (Figures 1 and 2). The low percentage of viable melanocytes could be related to the fact that melanocytes are target cells for UV toxicity by acting as shields for the nuclei and for the other skin cells [54]. During UV irradiation, melanin suffers a photosensitization process that results in the production of reactive oxygen species and the lethal insult of individual cells [30]. The susceptibility of murine epidermis skin (JB6 Cl 41-5a) cells to UVB irradiation could be ascribed to the fact that these cells are isolated from primary cultures of neonatal BALB/c epidermal cells, the newborn mice being the most suitable animal model to develop UV-induced melanoma [12]. Similar results regarding the noxious effect of UVB radiation on keratinocytes and fibroblasts viability were described in other studies, the intensity of the cytotoxic effect being dependent on the UVB dose and the experimental conditions applied [55–59]. A recent study showed that a higher dose of UVB (70 mJ/cm^2) used was nontoxic for fibroblasts [60]. It is well-known that UV radiation affects human skin at physiological, biological, and molecular levels by generating reactive oxygen species that are responsible for DNA damage, cell cycle arrest, and apoptosis, together with increased matrix metalloproteinase and elastase expression, having as a consequence, wrinkle formation and impaired cell migration [57].

UVB irradiation of healthy cells (HaCaT, 1BR3, HEMa, and JB6 Cl 41-5a), followed by hormone stimulation, led to some interesting results concerning their viability status: an increased viability percentage was recorded in all cell lines after UVB irradiation + EE or LNG (at both 1 and 10 µM) as compared to control UVB-irradiated cells (like the hormones "helped" the cells to recover after UVB damage), whereas UVB irradiation + EE + LNG (at 10 µM) proved to be toxic for all cells, and still less toxic as compared to UVB-irradiated cells (see Figures 1 and 2). The morphological features of the healthy cells changed significantly after UVB irradiation (Figures 6–9), with data that are confirmed by other studies in the literature [50,57,58]. The cells also stimulated with test hormones showed a lesser extent of damage; most of them presented characteristics similar with the control unexposed cells, results consistent with the viability data.

Considering the increased interest assigned to a possible link between sex hormones/oral contraceptive use and the development of melanoma, and the gaps existent in this regard, we assessed the impact of EE, LNG, and EE + LNG ± UVB irradiation on human (A375) and murine (B164A5) melanoma cells to provide reliable data concerning the current controversial reports. The test hormones exerted a dose-dependent cytotoxic effect on both A375 and B164A5 melanoma cells, the lowest percentage of viable cells being recorded after stimulation with EE + LNG—10 µM (56% and 47.23%, respectively) (see Figure 3). A considerable number of cells were floating, and this observation determined us to verify the type of cell death induced by the test compounds. An annexin V/PI test confirmed that the test hormones induced apoptosis of melanoma cells, the proapoptotic effect was also dose-dependent, and the strongest activity was triggered by EE + LNG (see Figures 4 and 5). The choice of the two different melanoma cell lines—A375 (human amelanotic cells) and B164A5 (murine melanotic cells) was based on the different response that was recorded in terms of melanoma aggressiveness, overall survival, and anti-melanoma therapies [30–34], our results being in accordance with these data.

Moroni and collaborators showed that low concentrations of progesterone (from 0.01 up to 1 µM) stimulate A375 melanoma cells proliferation, whereas higher concentrations (10–1000 µM) induce cell density reduction as a result of both cell cycle arrest and apoptosis [40]. Progesterone elicited a dose-dependent inhibitory effect on human melanoma (BLM) cell growth in vitro by inducing autophagy, but estrogen had no inhibitory effect [49]. A similar inhibitory activity of progesterone was observed in mouse melanoma cells—B16F10 [49]. The role of estrogens in melanoma susceptibility and malignancy remained controversial, due to the reported contradictory experimental and clinical findings: estradiol enhances tumor growth and metastasis in B16 melanoma cells, but in human malignant melanoma biopsies, the expressions of estrogen receptors ERα and ERβ are decreased [14]. Several studies described a suppressive role of 17β-estradiol on human SK-Mel-23 melanoma cell (these cells express only ERβ) proliferation [61]. An anti-invasive effect of 17β-estradiol was described in human melanoma cells devoid of ERα receptor [9]. A metabolite of estradiol, 2-methoxyestradiol proved in vitro and in vivo antimelanoma activity [9]. The morphology of A375 and B164A5 melanoma cells following hormone stimulation (1 µM) suffered several changes featured by the round shape of the floating cells that entered apoptosis, whereas the unaffected cells were strongly adherent and similar in shape to the control cells (Figures 10 and 11).

If the cytotoxic profile of test hormones (EE, LNG, and EE + LNG) was similar in human (A375) and murine (B164A5) melanoma cells, the intervention of UVB irradiation determined a different outcome, as follows: (i) in the case of A375 cells, the viability kept the same pattern as after hormones stimulation—a significantly reduced percentage of viable cells (dose-dependent) (Figure 3) and an increased percentage of proapoptotic cells (Figure 4) as compared to UVB-irradiated cells, the strongest effect being recorded for EE + LNG + UVB cells; and (ii) B164A5 cells viability was affected by UVB radiation, but the association of UVB and test hormones led to a lesser cytotoxic effect (Figure 3) and a lower percentage of apoptotic cells (Figure 5) as compared to hormone-only cell stimulation, as UVB made these cells more resistant to hormone cytotoxicity.

The differences regarding the behavior of A375 and B164A5 melanoma cells after UVB radiation could be attributed to the biological features of each cell line, in terms of: (i) origin: A375—human melanoma cells and B164A5—murine melanoma cells; (ii) morphology: A375 cells present an epithelial morphology and a reduced capacity to determine metastasis, whereas B164A5 cells have a fibroblastic-like morphology and are highly invasive/metastatic, and (iii) melanin content: A375 cells are devoid of melanin while B164A5 cells are melanin-producing cells [62].

Several studies reported a decrease of B16 melanoma cells viability after UVB radiation dependent on the UVB dose [63,64], data that are consistent with our results. Another possible explanation for B164A5 melanoma cells behavior in response to UVB irradiation and hormonal stimulation could be related to melanin, the pigment that is abundantly produced by B164A5 cells. A recent study highlighted the differences at the transcriptomic level between keratinocytes and melanocytes (main UV radiation targets in the skin), melanin representing a key player in the resistance/protection of melanocytes against UVB-induced damage. Melanin is able to counteract the acute effects of UVB radiation on melanocytes by absorbing the radiation. Moreover, it was stated that UV irradiation determined the chemiexcitation of melanin characterized by a continuous release of excited electrons, which has as consequence, DNA-damaged melanocytes long after UV exposure. These data underline an increased susceptibility of keratinocytes to UVB radiation in terms of toxicity as compared to melanocytes [59].

Another mechanism for melanocytes protection against UVB damage or carcinogenesis consists of the development of melanocytic dendrites that act as transporters of melanin pigment from melanocytes to neighboring keratinocytes in response to UVB radiation and hormonal treatment. A similar process of growing dendrites was described in melanoma cells after UV irradiation. Exposure of B16 melanoma cells to a dose of 100 mJ/cm^2 UVB led to morphological modifications of the cells, characterized by apparition of globular cell bodies and a high number of tree branch-like dendrites [64]. Based on these considerations, we could assume that UVB exposure, together with hormones stimulation of B164A5 melanoma cells led to an increased production of melanin, and to the apparition of dendrites reversing; therefore the cytotoxic effects exerted by the tested hormones in the absence of UVB irradiation, but this hypothesis needs to be further verified. This kind of effect was not observed in A375 melanoma cells due to the lack of melanin in these cells composition.

UVB irradiation ± hormonal treatment induced modifications of melanoma cells morphology (Figures 10 and 11): A375 cells: reduced confluence, round, detached, and floating apoptotic cells; B164A5 cells: round cells, cell shrinkage, and the presence of dendrites. Our data agree with the ones described in the literature that demonstrated that UVB irradiation induced the reorganization of cytoskeletal F-actin with globular cell bodies and a high number of dendrites in B16 melanoma cells [64]. In the presence of hormonal treatment, B164A5 cells began to recover their initial shape (Figure 11), an effect that was also observed in human dermal fibroblasts after stimulation with estrogen [47].

The test hormones were also investigated by an in ovo method to assess qualitatively an irritancy potential after topical application on mucosal or skin tissues. The HET-CAM represents an optimal pre-screening alternative method before animal testing, which is also useful as safety assessment for cutaneous applications [65–67]. The evaluation was consonant with in vitro cytotoxic results for the samples unexposed to UVB radiation. LNG showed no irritation both at 1 and 10 µM in consistency, as also indicated by the in vitro low influence on the viability of keratinocytes (HaCaT) and fibroblasts (1BR3). EE alone induced the highest irritation only at the higher tested concentration of 10 µM, but still the effect was very weak compared to the positive control. EE at 1 µM can be considered as non-irritant. The combination of EE + LNG induced, as expected, an even weaker effect at 10 µM, and no irritation at 1 µM. The test hormones are frequently used in micro-doses in transdermal systems or vaginal applications, and they are considered to be non-irritant [68,69]. Moreover, although EE and LNG are associated with vascular risk, in currently prescribed micro-doses does not induce endothelium–dependent vasodilatation [70]. Still, the evaluation of EE, LNG, and their combination in this chorioallantoic membrane environment, can be indicative for the effect on vascular modifications. This may explain why EE stimulates wound healing in in vitro keratinocytes and fibroblasts more than LNG, while, when studied in a vascular assay, LNG seem to attenuate EE effects on the capillary plexus.

4. Materials and Methods

4.1. Reagents and Cell Lines

Ethinylestradiol (EE) and levonorgestrel (LNG) analytical standards were acquired from Sigma Aldrich (Munich, Germany) and utilized as received. The test compounds (EE, LNG, and their combination—EE + LNG, in a molar ratio of 1:5) were dissolved in DMSO and were stored as stock solutions (5 mM) at 4 °C.

The experiment was conducted using four healthy and two tumor cell lines purchased as frozen items. The healthy cell lines, both human and murine, were as follows: HaCaT—immortalized human keratinocytes (ATCC, LGC Standards GmbH, Wesel, Germany), 1BR3—human skin fibroblast (90011801, ECACC General Collection, Salisbury, UK), HEMa—primary human epidermal melanocytes (ATCC, LGC Standards GmbH), and JB6Cl41-5a—newborn mice epidermis (CRL-2010™, ATCC, LGC Standards GmbH). The tumor cell lines, also human and murine, were: A375—human melanoma (CRL-1619™, ATCC, LGC Standards GmbH) and B164A5—murine melanoma (94042254; Sigma-Aldrich Chemie GmbH, Munich, Germany). All cell lines were kept in standard conditions before culture (liquid nitrogen).

The specific reagents for cell culture such as Dulbecco's modified Eagle's medium (DMEM), Eagle's Minimum Essential Medium (EMEM), Dermal Cell Basal Medium, and Adult Melanocyte Growth Kit were purchased from ATCC (LGC Standards GmbH); non-essential amino acids, fetal bovine serum (FBS), antibiotics mixture (penicillin/streptomycin), phosphate-buffered saline (PBS), trypsin/EDTA and Trypan blue were acquired from Sigma-Aldrich (Munich, Germany).

4.2. Cell Culture

Keratinocytes (HaCaT), and human (A375) and murine melanoma (B164A5) cell lines were cultured in DMEM high glucose (4.5 g/L) media, with 15 mM HEPES, and 2 mM L-glutamine supplemented with 10% FCS. A fibroblast (1BR3) cell line was cultured in EMEM supplemented with 15% FBS and for the mice epidermis (JB6 Cl 41-5a) cell line growth was used EMEM supplemented with 0.1% non-essential amino acids and 5% FCS. Primary melanocytes (HEMa) were grown in Dermal Cell Basal Medium supplemented with an Adult Melanocyte Growth Kit. An antibiotic mixture (100 U/mL penicillin, 100 µg/mL streptomycin) was added to all culture media, and the cells were preserved in standard conditions (humidified atmosphere with 5% CO_2 at 37 °C) and passaged every two days. Cell number was determined using a Countess II FL Automated Cell Counter (AMQAF1000, Thermo Fischer Scientific, Waltham, MA, USA) in the presence of Trypan blue. The cells were seeded in various culture plates (6, 12, and 96 wells) according to the experimental requirements.

4.3. UVB Irradiation Protocol

For UVB irradiation experiments, the cells were cultured in 6-/12- and 96-well plates, respectively, and allowed to grow until a confluence of 80–85% was achieved. The protocol consisted of several steps, as follows: the medium was removed prior to UVB exposure to avoid the formation of toxic photoproducts released by the medium [71], and the cells were washed with PBS (phosphate saline buffer); a thin layer of PBS was added in each well. UVB exposure was performed at 312 nm, at a dose of 40 mJ/cm^2 by means of Biospectra system (Vilber Lourmat, France). Immediately after irradiation, PBS was replaced with culture medium ± test compounds. The stimulation with test compounds (LNG, EE, and EE + LNG) was performed after UVB irradiation.

4.4. Cell Viability, Migration and Proliferation Assays

Viability assessment. The viability test applied in the current study was the Alamar blue assay. The cells (1 × 10^4/200 µL medium/well) were seeded in a 96-well plate and allowed to attach; afterwards, were incubated with different concentrations (1 and 10 µM) of test compounds for 24 h. The absorbance was measured using a xMark™ Microplate Spectrophotometer (BioRad) at 570 nm and

600 nm (reference) wavelengths; and cell viability was calculated according to the method described in our previous studies [72].

Migration and proliferation assay. The migratory character of the cells used in the present study was evaluated by means of a scratch assay, a wound healing type technique. In brief, a number of 2×10^5 cells/well were cultured in 12-well plates, and when the suitable confluence (~90–95%) was reached, a scratch was performed in the middle of the well with a 10 μL sterile tip [73]. To quantify the effect of the test compounds (1 μM EE, LNG, and EE + LNG, respectively) in terms of cell migratory capacity, the difference between the initial and after 24 h wound widths, was determined. Representative images (10× magnification) were recorded by using an Olympus IX73 inverted microscope equipped with DP74 camera (Olympus, Tokyo, Japan) and the wound widths were measured with CellSense Dimension 1.17 (Olympus, Tokyo, Japan). The migration rate was calculated according to the formula described by Felice et al. [74].

Annexin V/PI assay. In order to study the impact of test compounds on cell apoptosis, flow cytometry analysis was performed using an annexin V-FITC apoptosis detection kit (eBioscience, Vienna, Austria). A375 human melanoma and B164A5 murine melanoma cells were seeded into 6-well plates (3×10^5 cells/well) and stimulated with test compounds (1 and 10 μM) for 24 h. After 24 h, the cells were washed with PBS and resuspended in 200 μL Binding Buffer; 5 μL of FITC-conjugated annexin V were added into the cell suspension. Before analysis, 10 μL of propidium iodide solution (20 μg/mL) were added in each sample, followed by 10 min incubation at room temperature in the dark. Cells were analyzed by flow cytometry (FACSCalibur; Becton Dickson, Franklin Lakes, NJ, USA) and unstimulated cells were used as controls. The results were processed using Flowing Software Version 2.5.1 (developed by Perttu Terho, Cell Imaging Core, Turku Centre for Biotechnology, Turun Yliopisto, Finland).

4.5. Hen's Egg Test—Chorioallantoic Membrane (HET-CAM) Assay

The evaluation of hormones biocompatibility and toxicity was assessed in ovo by the Hen's Egg Chorioallantoic Membrane Test (HET-CAM). The method is applied to evaluate a potential irritant effect of the test compounds on the vascular plexus of the chorioallantoic membrane [65,75]. The HET-CAM method was carried out following ICCVAM recommendations and adapted to our conditions [76]. Thus, the eggs were incubated at 37 °C and controlled humidity. On the third day of incubation (embryonic day of development, EDD 3), 3–4 mL of albumen were extracted in order to facilitate the observation of the chorioallantoic membrane: a hole was cut in the lower part of the egg, which was then covered, and the eggs were reintroduced into the incubator. On EDD 4, a window was cut and removed from the top of the eggs. The hole was then covered, and the eggs were kept in the incubator until EDD 9. Five eggs were used for each tested compound. A volume of 500 μL of control or test solution, respectively, was applied and the modifications produced at the CAM level were monitored by means of stereomicroscopy (Discovery 8 Stereomicroscope, Zeiss, Göttingen, Germany); significant images were recorded (Axio CAM 105 color, Zeiss) before the application and after 5 min of contact with each sample. All images were processed using AxioVision SE64. Rel. 4.9.1 Software (Zeiss), Gimp v 2.8 (https://www.gimp.org/) and ImageJ v 1.50e software (U.S. National Institutes of Health, Bethesda, MD, USA).

The negative control was represented by a phosphate buffer solution (PBS), while the positive control by the sodium dodecyl sulfate (SDS) 1% in PBS. The test compounds were diluted in DMSO at concentrations of 1 μM and 10 μM.

The time needed for the test compounds to induce a particular reaction (hemorrhage—H—blood vessel bleeding, vascular lysis—L—disintegration of blood vessels, coagulation—C—intra or extra-vascular protein denaturizing) was recorded in seconds and was established at 5 min (300 s). The analytical method used to assess the irritant potential of test compounds consisted in calculating the irritation score (IS), using the formula described in our previous study [77]. The formula comprises a factor indicating the impact on vascular damage of the observed effect, e.g., coagulation has the highest impact on irritancy, being represented by a multiplication factor of 9. Therefore, the irritation scores may have values between 0 and 21 [75].

To establish the irritation severity, a severity score (SS) was also calculated. After 5 min of observation, the most pronounced reaction was scored (either hemorrhage, lysis, or coagulation) according to the following scheme: 0 = no reaction; 1 = slight reaction; 2 = moderate reaction; 3 = severe reaction. Mean scores were determined.

4.6. Statistical Analysis

The statistical program and software applied in the present study were GraphPad Prism 7 (GraphPad Software, La Jolla, CA, USA), CellSense Dimension 1.17 software, Flowing Software Version 2.5.1, AxioVision SE64. Rel. 4.9.1 Software, Gimp v 2.8 and ImageJ v 1.50e software. Data were analyzed using paired Student's *t* tests or one-way ANOVA, followed by Tukey's post-tests when appropriate, to determine the statistical difference between experimental and control groups; *, **, *** and **** indicate $p < 0.05$, $p < 0.01$, $p < 0.001$ and $p < 0.0001$, respectively.

5. Conclusions

This initial study that evaluates the potential antimelanoma activity of ethinylestradiol, levonorgestrel, and their combination ± UVB irradiation—proposed an approach that might prove its utility in further studies. The two hormones (EE and LNG) and their combination (EE + LNG) did not interfere with human and murine healthy skin cell viability at the lowest concentration tested, whereas in the case of melanoma cells, this concentration induced a significant cytotoxic effect. By increasing the concentration of hormones and adding UVB irradiation, the cytotoxicity induced was at a higher extent in all healthy cell lines and in human melanoma cells—A375. In the case of murine melanoma cells—B164A5, the association of hormones and UVB stress led to an increase of viable cell percentage and a decrease of early apoptotic cells, a possible key role being played by melanin. In ovo experiments confirmed the harmless activity of the hormones at low doses, albeit a higher concentration was responsible for a weak irritant effect (for EE and EE + LNG). These experimental observations offer a reliable background for further in vitro studies to define the mechanisms involved in cell type-dependent antimelanoma activity exerted by the two hormones, and to explain the possible role of melanin in the protection of melanoma cells against hormonal treatment.

Supplementary Materials: The supplementary materials are available online at http://www.mdpi.com/1422-0067/19/11/3600/s1.

Author Contributions: Conceptualization, D.C., I.P. and C.D.; Funding acquisition, R.A.P.; Investigation, S.S. and C.F.; Methodology, C.N.; Project administration, C.N.; Resources, D.S. and C.D.; Software, D.N.; Supervision, C.D.; Validation, D.C., I.P., C.S., and C.D.; Visualization, C.S.; Writing—original draft, M.P. and S.A.; Writing—review & editing, D.C., C.F., and I.P.

Funding: This research was funded by the internal grants PIII-C4-PCFI-2016/2017-04 and PIII-C5-PCFI-2017/2018-04 offered by the "Victor Babes" University of Medicine and Pharmacy (S.A.).

Acknowledgments: The experiments were conducted within the Center of Pharmaco-toxicological evaluations from the Faculty of Pharmacy, "Victor Babes" University of Medicine and Pharmacy, Timisoara.

Conflicts of Interest: The authors declare no conflict of interest.

Abbreviations

A375	Human melanoma cells
B164A5	Murine melanoma cells
BCC	Basal cell carcinoma
1BR3	Human skin fibroblasts
DMSO	Dimethyl sulfoxide
EE	Ethinylestradiol
ER	Estrogen receptor
ET-1	Endothelin-1
HaCaT	Human immortalized keratinocytes
HEMa	Human primary epidermal melanocytes
HET-CAM	Hen's Egg Test—chick chorioallantoic membrane assay
ICCVAM	Interagency Coordinating Committee on the Validation of Alternative Methods
IS	Irritation score
JB6Cl415a	Newborn mice epidermis
LNG	Levonorgestrel
MSH	Melanocyte-stimulating hormone
PBS	Phosphate saline buffer
SCC	Squamous cell carcinoma
SCF	Stem cell factor
SDS	Sodium dodecyl sulfate
SS	Severity score
UVB	Ultraviolet B radiation
UVA	Ultraviolet A radiation

References

1. Burkman, R.; Schlesselman, J.J.; Zieman, M. Safety concerns and health benefits associated with oral contraception. *Am. J. Obstet. Gynecol.* **2004**, *190* (Suppl. 4), S5–S22. [CrossRef] [PubMed]
2. Stanczyk, F.Z.; Archer, D.F.; Bhavnani, B.R. Ethinyl estradiol and 17β-estradiol in combined oral contraceptives: Pharmacokinetics, pharmacodynamics and risk assessment. *Contraception* **2013**, *87*, 706–727. [CrossRef] [PubMed]
3. Leslie, K.K.; Espey, E. Oral contraceptives and skin cancer: Is there a link? *Am. J. Clin. Dermatol.* **2005**, *6*, 349–355. [CrossRef] [PubMed]
4. Dhont, M. History of oral contraception. *Eur. J. Contracept. Reprod. Health Care* **2010**, *15* (Suppl. 2), S12–S18. [CrossRef] [PubMed]
5. Lata, K.; Mukherjee, T.K. Knockdown of receptor for advanced glycation end products attenuate 17α-ethinyl-estradiol dependent proliferation and survival of MCF-7 breast cancer cells. *Biochim. Biophys. Acta* **2014**, *1840*, 1083–1091. [CrossRef] [PubMed]
6. Kim, S.K.; Shin, S.J.; Yoo, Y.; Kim, N.H.; Kim, D.S.; Zhang, D.; Park, J.A.; Yi, H.; Kim, J.S.; Shin, H.C. Oral toxicity of isotretinoin, misoprostol, methotrexate, mifepristone and levonorgestrel as pregnancy category X medications in female mice. *Exp. Ther. Med.* **2015**, *9*, 853–859. [CrossRef] [PubMed]
7. Iversen, L.; Sivasubramaniam, S.; Lee, A.J.; Fielding, S.; Hannaford, P.C. Lifetime cancer risk and combined oral contraceptives: The Royal College of General Practitioners' Oral Contraception Study. *Am. J. Obstet. Gynecol.* **2017**, *216*, 580.e1–580.e9. [CrossRef] [PubMed]
8. Janik, M.E.; Bełkot, K.; Przybyło, M. Is oestrogen an important player in melanoma progression? *Contemp. Oncol. (Pozn)* **2014**, *18*, 302–306. [CrossRef] [PubMed]
9. Marzagalli, M.; Montagnani Marelli, M.; Casati, L.; Fontana, F.; Moretti, R.M.; Limonta, P. Estrogen Receptor β in Melanoma: From Molecular Insights to Potential Clinical Utility. *Front. Endocrinol. (Lausanne)* **2016**, *7*, 140. [CrossRef] [PubMed]
10. Dika, E.; Fanti, P.A.; Vaccari, S.; Capizzi, E.; Degiovanni, A.; Gobbi, A.; Piraccini, B.M.; Ribero, S.; Baraldi, C.; Ravaioli, G.M.; et al. Oestrogen and progesterone receptors in melanoma and nevi: An immunohistochemical study. *Eur. J. Dermatol.* **2017**, *27*, 254–259. [CrossRef] [PubMed]

11. Kuklinski, L.F.; Zens, M.S.; Perry, A.E.; Gossai, A.; Nelson, H.H.; Karagas, M.R. Sex hormones and the risk of keratinocyte cancers among women in the United States: A population-based case-control study. *Int. J. Cancer* **2016**, *139*, 300–309. [CrossRef] [PubMed]

12. Coricovac, D.; Dehelean, C.; Moaca, E.A.; Pinzaru, I.; Bratu, T.; Navolan, D.; Boruga, O. Cutaneous Melanoma—A Long Road from Experimental Models to Clinical Outcome: A Review. *Int. J. Mol. Sci.* **2018**, *19*, 1566. [CrossRef] [PubMed]

13. Roh, M.R.; Eliades, P.; Gupta, S.; Grant-Kels, J.M.; Tsao, H. Cutaneous melanoma in women. *Int. J. Women's Dermatol.* **2017**, *3* (Suppl. 1), S11–S15. [CrossRef] [PubMed]

14. Mitchell, D.L.; Fernandez, A.A.; Garcia, R.; Paniker, L.; Lin, K.; Hanninen, A.; Zigelsky, K.; May, M.; Nuttall, M.; Lo, H.H.; et al. Acute exposure to ultraviolet-B radiation modulates sex steroid hormones and receptor expression in the skin and may contribute to the sex bias of melanoma in a fish model. *Pigment Cell Melanoma Res.* **2014**, *27*, 408–417. [CrossRef] [PubMed]

15. Tchernev, G.; Dzhelyatova, G.A.; Wollina, U.; Lozev, I.; Lotti, T. Medium Sized Congenital Melanocytic Nevus with Suspected Progression to Melanoma during Pregnancy: What's the Best for the Patient? *J. Med. Sci.* **2018**, *6*, 143–145. [CrossRef] [PubMed]

16. Smalley, K.S. Why do women with melanoma do better than men? *eLife* **2018**, *7*, E33511. [CrossRef] [PubMed]

17. Tamega Ade, A.; Miot, H.A.; Moço, N.P.; Silva, M.G.; Marques, M.E.; Miot, L.D. Gene and protein expression of oestrogen-β and progesterone receptors in facial melasma and adjacent healthy skin in women. *Int. J. Cosmet. Sci.* **2015**, *37*, 222–228. [CrossRef] [PubMed]

18. Cohen, P.R. Melasma treatment: A novel approach using a topical agent that contains an anti-estrogen and a vascular endothelial growth factor inhibitor. *Med. Hypotheses* **2017**, *101*, 1–5. [CrossRef] [PubMed]

19. De Giorgi, V.; Gori, A.; Grazzini, M.; Rossari, S.; Scarfi, F.; Corciova, S.; Verdelli, A.; Lotti, T.; Massi, D. Estrogens, estrogen receptors and melanoma. *Expert Rev. Anticancer Ther.* **2011**, *11*, 739–747. [CrossRef] [PubMed]

20. Pelletier, G.; Ren, L. Localization of sex steroid receptors in human skin. *Histol. Histopathol.* **2004**, *19*, 629–636. [CrossRef] [PubMed]

21. Botteri, E.; Støer, N.C.; Sakshaug, S.; Graff-Iversen, S.; Vangen, S.; Hofvind, S.; Ursin, G.; Weiderpass, E. Menopausal hormone therapy and risk of melanoma: Do estrogens and progestins have a different role? *Int. J. Cancer* **2017**, *141*, 1763–1770. [CrossRef] [PubMed]

22. Natale, C.A.; Duperret, E.K.; Zhang, J.; Sadeghi, R.; Dahal, A.; O'Brien, K.T.; Cookson, R.; Winkler, J.D.; Ridky, T.W. Sex steroids regulate skin pigmentation through nonclassical membrane-bound receptors. *eLife* **2016**, *5*, E15104. [CrossRef] [PubMed]

23. Rastrelli, M.; Tropea, S.; Rossi, C.R.; Alaibac, M. Melanoma: Epidemiology, risk factors, pathogenesis, diagnosis and classification. *In Vivo* **2014**, *28*, 1005–1011. [PubMed]

24. Leslie, K.S.; Lodge, E.; Garioch, J.J. A comparison of narrowband (TL-01) UVB-induced erythemal response at different body sites. *Clin. Exp. Dermatol.* **2005**, *30*, 337–339. [CrossRef] [PubMed]

25. Ehrlich, M. DNA hypomethylation in cancer cells. *Epigenomics* **2009**, *1*, 239–259. [CrossRef] [PubMed]

26. Richarz, N.A.; Aguilera, J.; Castillo, G.; Fuente, M.J.; Ferrándiz, C.; Carrascosa, J.M. Phototoxic reaction to a combined oral contraceptive (levonorgestrel/ethinylestradiol). *Photochem. Photobiol. Sci.* **2017**, *16*, 1381–1383. [CrossRef] [PubMed]

27. Yoon, H.S.; Shin, C.Y.; Kim, Y.K.; Lee, S.R.; Chung, J.H. Endogenous estrogen exacerbates UV-induced inflammation and photoaging in mice. *J. Investig. Dermatol.* **2014**, *134*, 2290–2293. [CrossRef] [PubMed]

28. Yoon, H.S.; Chung, J.H. Long-Term Estrogen Effects on Sun-Exposed Human Skin in M.A. In *Textbook of Aging Skin*; Farage, M., Miller, K., Maibach, H., Eds.; Springer: Berlin/Heidelberg, Germany, 2015.

29. Röck, K.; Joosse, S.A.; Müller, J.; Heinisch, N.; Fuchs, N.; Meusch, M.; Zipper, P.; Reifenberger, J.; Pantel, K.; Fischer, J.W. Chronic UVB-irradiation actuates perpetuated dermal matrix remodeling in female mice: Protective role of estrogen. *Sci. Rep.* **2016**, *6*, 30482. [CrossRef] [PubMed]

30. Slominski, A.; Tobin, D.J.; Shibahara, S.; Wortsman, J. Melanin pigmentation in mammalian skin and its hormonal regulation. *Physiol. Rev.* **2004**, *84*, 1155–1228. [CrossRef] [PubMed]

31. Slominski, A.; Kim, T.K.; Brożyna, A.A.; Janjetovic, Z.; Brooks, D.L.; Schwab, L.P.; Skobowiat, C.; Jóźwicki, W.; Seagroves, T.N. The role of melanogenesis in regulation of melanoma behavior: Melanogenesis leads to stimulation of HIF-1α expression and HIF-dependent attendant pathways. *Arch. Biochem. Biophys.* **2014**, *563*, 79–93. [CrossRef] [PubMed]

32. Slominski, R.M.; Zmijewski, M.A.; Slominski, A.T. The role of melanin pigment in melanoma. *Exp. Dermatol.* **2015**, *24*, 258–259. [CrossRef] [PubMed]

33. Brożyna, A.A.; Jóźwicki, W.; Roszkowski, K.; Filipiak, J.; Slominski, A.T. Melanin content in melanoma metastases affects the outcome of radiotherapy. *Oncotarget* **2016**, *7*, 17844–17853. [CrossRef] [PubMed]

34. Brożyna, A.A.; Jóźwicki, W.; Carlson, J.A.; Slominski, A.T. Melanogenesis affects overall and disease-free survival in patients with stage III and IV melanoma. *Hum. Pathol.* **2013**, *44*, 2071–2074. [CrossRef] [PubMed]

35. Interagency Coordinating Committee on the Validation of Alternative Methods (ICCVAM 2010). Available online: https://www.niehs.nih.gov/health/materials/interagency_coordinating_committee_on_the_validation_of_alternative_methods_508.pdf (accessed on 1 August 2018).

36. Gallagher, R.P.; Elwood, J.M.; Hill, G.B.; Coldman, A.J.; Threlfall, W.J.; Spinelli, J.J. Reproductive factors, oral contraceptives and risk of malignant melanoma: Western Canada Melanoma Study. *Br. J. Cancer* **1985**, *52*, 901–907. [CrossRef] [PubMed]

37. Holman, C.D.; Armstrong, B.K.; Heenan, P.J. Cutaneous malignant melanoma in women: Exogenous sex hormones and reproductive factors. *Br. J. Cancer* **1984**, *50*, 673–680. [CrossRef] [PubMed]

38. Smith, M.A.; Fine, J.A.; Barnhill, R.L.; Berwick, M. Hormonal and reproductive influences and risk of melanoma in women. *Int. J. Epidemiol.* **1998**, *27*, 751–757. [CrossRef] [PubMed]

39. Koomen, E.R.; Joosse, A.; Herings, R.M.; Casparie, M.K.; Guchelaar, H.J.; Nijsten, T. Estrogens, oral contraceptives and hormonal replacement therapy increase the incidence of cutaneous melanoma: A population-based case-control study. *Ann. Oncol.* **2009**, *20*, 358–364. [CrossRef] [PubMed]

40. Moroni, G.; Gaziano, R.; Bue, C.; Agostini, M.; Perno, C.F.; Sinibaldi-Vallebona, P.; Pica, F. Progesterone and Melanoma Cells: An Old Story Suspended between Life and Death. *J. Steroids Horm. Sci.* **2015**, *S13*, 158. [CrossRef]

41. Karagas, M.R.; Stukel, T.A.; Dykes, J.; Miglionico, J.; Greene, M.A.; Carey, M.; Armstrong, B.; Elwood, J.M.; Gallagher, R.P.; Green, A.; et al. A pooled analysis of 10 case-control studies of melanoma and oral contraceptive use. *Br. J. Cancer* **2002**, *86*, 1085–1092. [CrossRef] [PubMed]

42. Mueller, K.; Verzi, A.E.; Bhatt, K.; Orrell, K.; Hagstorm, E.; Flood, K.; Schlosser, B.; Nardone, B.; West, D.P. Melanoma and chronic exposure to contraceptives containing microdoses of ethinylestradiol in young women: A retrospective study from the Research on Adverse Drug Events and Reports (RADAR) project comprising a large Midwestern U.S. patient population. *J. Eur. Acad. Dermatol. Venereol.* **2018**, *32*, e87–e88. [CrossRef] [PubMed]

43. Bhupathiraju, S.N.; Grodstein, F.; Stampfer, M.J.; Willett, W.C.; Hu, F.B.; Manson, J.E. Exogenous Hormone Use: Oral Contraceptives, Postmenopausal Hormone Therapy, and Health Outcomes in the Nurses' Health Study. *Am. J. Public Health* **2016**, *106*, 1631–1637. [CrossRef] [PubMed]

44. Mordoh, J.; Tapia, I.J.; Barrio, M.M. A word of caution: Do not wake sleeping dogs; micrometastases of melanoma suddenly grew after progesterone treatment. *BMC Cancer* **2013**, *13*, 132. [CrossRef] [PubMed]

45. Caraffa, A.; Spinas, E.; Kritas, S.K.; Lessiani, G.; Ronconi, G.; Saggini, A.; Antinolfi, P.; Pizzicannella, J.; Toniato, E.; Theoharides, T.C.; et al. Endocrinology of the skin: Intradermal neuroimmune network, a new frontier. *J. Biol. Regul. Homeost. Agents* **2016**, *30*, 339–343. [PubMed]

46. Muizzuddin, N.; Marenus, K.D.; Schnittger, S.F.; Sullivan, M.; Maes, D.H. Effect of systemic hormonal cyclicity on skin. *J. Cosmet. Sci.* **2005**, *56*, 311–321. [CrossRef] [PubMed]

47. Carnesecchi, J.; Malbouyres, M.; de Mets, R.; Balland, M.; Beauchef, G.; Vié, K.; Chamot, C.; Lionnet, C.; Ruggiero, F.; Vanacker, J.M. Estrogens induce rapid cytoskeleton re-organization in human dermal fibroblasts via the non-classical receptor GPR30. *PLoS ONE* **2015**, *10*, e0120672. [CrossRef] [PubMed]

48. Natale, C.A.; Li, J.; Zhang, J.; Dahal, A.; Dentchev, T.; Stanger, B.Z.; Ridky, T.W. Activation of G protein-coupled estrogen receptor signaling inhibits melanoma and improves response to immune checkpoint blockade. *eLife* **2018**, *7*, E31770. [CrossRef] [PubMed]

49. Ramaraj, P.; Cox, J.L. In vitro effect of progesterone on human melanoma (BLM) cell growth. *Int. J. Clin. Exp. Med.* **2014**, *7*, 3941–3953. [PubMed]

50. Fernandez, T.L.; Van Lonkhuyzen, D.R.; Dawson, R.A.; Kimlin, M.G.; Upton, Z. In vitro investigations on the effect of dermal fibroblasts on keratinocyte responses to ultraviolet B radiation. *Photochem. Photobiol.* **2014**, *90*, 1332–1339. [CrossRef] [PubMed]

51. Duval, C.; Cohen, C.; Chagnoleau, C.; Flouret, V.; Bourreau, E.; Bernerd, F. Key Regulatory Role of Dermal Fibroblasts in Pigmentation as Demonstrated Using a Reconstructed Skin Model: Impact of Photo-Aging. *PLoS ONE* **2014**, *9*, e114182. [CrossRef] [PubMed]

52. Wiedemann, C.; Nägele, U.; Schramm, G.; Berking, C. Inhibitory effects of progestogens on the estrogen stimulation of melanocytes in vitro. *Contraception* **2009**, *80*, 292–298. [CrossRef] [PubMed]

53. Kanda, N.; Watanabe, S. 17beta-estradiol stimulates the growth of human keratinocytes by inducing cyclin D2 expression. *J. Investig. Dermatol.* **2004**, *123*, 319–328. [CrossRef] [PubMed]

54. Poletini, M.O.; de Assis, L.V.; Moraes, M.N.; Castrucci, A.M. Estradiol differently affects melanin synthesis of malignant and normal melanocytes: A relationship with clock and clock-controlled genes. *Mol. Cell Biochem.* **2016**, *421*, 29–39. [CrossRef] [PubMed]

55. Izykowska, I.; Cegielski, M.; Gebarowska, E.; Podhorska-Okolow, M.; Piotrowska, A.; Zabel, M.; Dziegiel, P. Effect of melatonin on human keratinocytes and fibroblasts subjected to UVA and UVB radiation In vitro. *In Vivo* **2009**, *23*, 739–745. [PubMed]

56. Mortensen, L.J.; Ravichandran, S.; DeLouise, L.A. The impact of UVB exposure and differentiation state of primary keratinocytes on their interaction with quantum dots. *Nanotoxicology* **2013**, *7*, 1244–1254. [CrossRef] [PubMed]

57. Lee, G.T.; Cha, H.J.; Lee, K.S.; Lee, K.K.; Hong, J.T.; Ahn, K.J.; An, I.S.; An, S.; Bae, S. Arctiin induces an UVB protective effect in human dermal fibroblast cells through microRNA expression changes. *Int. J. Mol. Med.* **2014**, *33*, 640–648. [CrossRef] [PubMed]

58. Patwardhan, J.; Bhatt, P. Ultraviolet-B Protective Effect of Flavonoids from Eugenia caryophylata on Human Dermal Fibroblast Cells. *Pharmacogn. Mag.* **2015**, *11* (Suppl. 3), S397–S406. [CrossRef] [PubMed]

59. Sun, X.; Kim, A.; Nakatani, M.; Shen, Y.; Liu, L. Distinctive molecular responses to ultraviolet radiation between keratinocytes and melanocytes. *Exp. Dermatol.* **2016**, *25*, 708–713. [CrossRef] [PubMed]

60. Shin, D.; Lee, S.; Huang, Y.H.; Lim, H.W.; Lee, Y.; Jang, K.; Cho, Y.; Park, S.J.; Kim, D.D.; Lim, C.J. Protective properties of geniposide against UV-B-induced photooxidative stress in human dermal fibroblasts. *Pharm. Biol.* **2018**, *56*, 176–182. [CrossRef] [PubMed]

61. Sarti, M.S.; Visconti, M.A.; Castrucci, A.M. Biological activity and binding of estradiol to SK-Mel 23 human melanoma cells. *Braz. J. Med. Biol. Res.* **2004**, *37*, 901–905. [CrossRef] [PubMed]

62. Serafim, V.; Shah, A.; Puiu, M.; Andreescu, N.; Coricovac, D.; Nosyrev, A.; Spandidos, D.A.; Tsatsakis, A.M.; Dehelean, C.; Pinzaru, I. Classification of cancer cell lines using matrix-assisted laser desorption/ionization time-of-flight mass spectrometry and statistical analysis. *Int. J. Mol. Med.* **2017**, *40*, 1096–1104. [CrossRef] [PubMed]

63. Jeong, Y.M.; Lee, J.E.; Kim, S.Y.; Yun, H.Y.; Baek, K.J.; Kwon, N.S.; Kim, D.S. Enhanced effects of citrate on UVB-induced apoptosis of B16 melanoma cells. *Pharmazie* **2009**, *64*, 829–833. [PubMed]

64. Wang, W.Q.; Wu, J.F.; Xiao, X.Q.; Xiao, Q.; Wang, J.; Zuo, F.G. Narrow-band UVB radiation promotes dendrite formation by activating Rac1 in B16 melanoma cells. *Mol. Clin. Oncol.* **2013**, *1*, 858–862. [CrossRef] [PubMed]

65. Wilson, T.D.; Steck, W.F. A modified HET-CAM assay approach to the assessment of anti-irritant properties of plant extracts. *Food Chem. Toxicol.* **2000**, *38*, 867–872. [CrossRef]

66. Rocha-Filho, P.; Ferrari, M.; Maruno, M.; Souza, O.; Gumiero, V. In Vitro and In Vivo Evaluation of Nanoemulsion Containing Vegetable Extracts. *Cosmetics* **2017**, *4*, 32. [CrossRef]

67. Palmeira-de-Oliveira, R.; Monteiro Machado, R.; Martinez-de-Oliveira, J.; Palmeira-de-Oliveira, A. Testing vaginal irritation with the hen's egg test-chorioallantoic membrane assay. *ALTEX* **2018**. [CrossRef] [PubMed]

68. Wieder, D.R.; Pattimakiel, L. Examining the efficacy, safety, and patient acceptability of the combined contraceptive vaginal ring (NuvaRing). *Int. J. Women's Health* **2010**, *2*, 401–409. [CrossRef] [PubMed]

69. Abdi, F.; Mobedi, H.; Mosaffa, N.; Dolatian, M.; Ramezani Tehrani, F. Hormone Therapy for Relieving Postmenopausal Vasomotor Symptoms: A Systematic Review. *Arch. Iran Med.* **2016**, *19*, 141–146. [PubMed]

70. Torgrimson, B.N.; Meendering, J.R.; Kaplan, P.F.; Minson, C.T. Endothelial function across an oral contraceptive cycle in women using levonorgestrel and ethinyl estradiol. *Am. J. Physiol. Heart Circ. Physiol.* **2007**, *292*, H2874–H2880. [CrossRef] [PubMed]

71. Lan, C.E.; Wang, Y.T.; Lu, C.Y.; Fang, A.H.; Wu, C.S. The effect of interaction of heat and UVB on human keratinocyte: Novel insights on UVB-induced carcinogenesis of the skin. *J. Dermatol. Sci.* **2017**, *88*, 207–215. [CrossRef] [PubMed]

72. Coricovac, D.E.; Moacă, E.A.; Pinzaru, I.; Cîtu, C.; Soica, C.; Mihali, C.V.; Păcurariu, C.; Tutelyan, V.A.; Tsatsakis, A.; Dehelean, C.A. Biocompatible Colloidal Suspensions Based on Magnetic Iron Oxide Nanoparticles: Synthesis, Characterization and Toxicological Profile. *Front. Pharmacol.* **2017**, *8*, 154. [CrossRef] [PubMed]

73. Jung, M.; Weigert, A.; Tausendschön, A.; Mora, J.; Oren, B.; Sola, A.; Hotter, G.; Muta, T.; Brüne, B. Interleukin-10-induced neutrophil gelatinase-associated lipocalin production in macrophages with consequences for tumor growth. *Mol. Cell Biol.* **2012**, *32*, 3938–3948. [CrossRef] [PubMed]

74. Felice, F.; Zambito, Y.; Belardinelli, E.; Fabiano, A.; Santoni, T.; Di Stefano, R. Effect of different chitosan derivatives on in vitro scratch wound assay: A comparative study. *Int. J. Biol. Macromol.* **2015**, *76*, 236–241. [CrossRef] [PubMed]

75. Luepke, N.P. Hen's egg chorioallantoic membrane test for irritation potential. *Food Chem. Toxicol.* **1985**, *23*, 287–291. [CrossRef]

76. Scheel, J.; Kleber, M.; Kreutz, J.; Lehringer, E.; Mehling, A.; Reisinger, K.; Steiling, W. Eye irritation potential: Usefulness of the HET-CAM under the Globally Harmonized System of classification and labeling of chemicals (GHS). *Regul. Toxicol. Pharmacol.* **2011**, *59*, 471–492. [CrossRef] [PubMed]

77. Ardelean, S.; Feflea, S.; Ionescu, D.; Năstase, V.; Dehelean, C.A. Toxicologic screening of some surfactants using modern in vivo bioassays. *Rev. Med. Chir. Soc. Med. Nat. Iasi* **2011**, *115*, 251–258. [PubMed]

MDPI

St. Alban-Anlage 66

4052 Basel

Switzerland

Tel. +41 61 683 77 34

Fax +41 61 302 89 18

www.mdpi.com

International Journal of Molecular Sciences Editorial Office

E-mail: ijms@mdpi.com

www.mdpi.com/journal/ijms